SpringerBriefs in Electrical and Computer Engineering

Control, Automation and Robotics

Series Editors

Tamer Başar, Coordinated Science Laboratory, University of Illinois at Urbana-Champaign, Urbana, IL, USA

Miroslav Krstic, La Jolla, CA, USA

SpringerBriefs in Control, Automation and Robotics presents concise summaries of theoretical research and practical applications. Featuring compact, authored volumes of 50 to 125 pages, the series covers a range of research, report and instructional content. Typical topics might include:

- a timely report of state-of-the art analytical techniques;
- a bridge between new research results published in journal articles and a contextual literature review;
- a novel development in control theory or state-of-the-art development in robotics;
- an in-depth case study or application example;
- a presentation of core concepts that students must understand in order to make independent contributions; or
- a summation/expansion of material presented at a recent workshop, symposium or keynote address.

SpringerBriefs in Control, Automation and Robotics allows authors to present their ideas and readers to absorb them with minimal time investment, and are published as part of Springer's e-Book collection, with millions of users worldwide. In addition, Briefs are available for individual print and electronic purchase.

Springer Briefs in a nutshell

- 50–125 published pages, including all tables, figures, and references;
- softcover binding;
- publication within 9–12 weeks after acceptance of complete manuscript;
- copyright is retained by author;
- authored titles only – no contributed titles; and
- versions in print, eBook, and MyCopy.

Indexed by Engineering Index.

Publishing Ethics: Researchers should conduct their research from research proposal to publication in line with best practices and codes of conduct of relevant professional bodies and/or national and international regulatory bodies. For more details on individual ethics matters please see: https://www.springer.com/gp/authors-editors/journal-author/journal-author-helpdesk/publishing-ethics/14214

More information about this subseries at http://www.springer.com/series/10198

Pierre-Jean Meyer · Alex Devonport ·
Murat Arcak

Interval Reachability Analysis

Bounding Trajectories of Uncertain Systems
with Boxes for Control and Verification

 Springer

Pierre-Jean Meyer
Department of EECS
University of California, Berkeley
Berkeley, CA, USA

Alex Devonport
Department of EECS
University of California, Berkeley
Berkeley, CA, USA

Murat Arcak
Department of EECS
University of California, Berkeley
Berkeley, CA, USA

ISSN 2191-8112   ISSN 2191-8120   (electronic)
SpringerBriefs in Electrical and Computer Engineering
ISSN 2192-6786   ISSN 2192-6794   (electronic)
SpringerBriefs in Control, Automation and Robotics
ISBN 978-3-030-65109-1   ISBN 978-3-030-65110-7   (eBook)
https://doi.org/10.1007/978-3-030-65110-7

MATLAB is a registered trademark of The MathWorks, Inc. See https://www.mathworks.com/ trademarks for a list of additional trademarks.

Mathematics Subject Classification: 93-01, 93B03, 93B40, 93C10, 93C15

This Springer imprint is published by the registered company Springer Nature Switzerland AG
The registered company address is: Gewerbestrasse 11, 6330 Cham, Switzerland

Preface

Reachability analysis deals with the problem of evaluating the set of all the successor states that can be reached in finite time by a system starting from a given set of initial states, and plays a key role in verifying or enforcing by control the satisfaction of safety-critical specifications. Although there exists a wide variety of methods and set representations to approximate reachable sets, this book focuses on computing reachable set over-approximations in the form of multi-dimensional intervals (hyperrectangles) due to their greater simplicity of use and lower computational complexity.

The main motivation and purpose of this book is to provide to the readers a tutorial presentation of several approaches for interval reachability analysis. Instead of focusing on mathematical developments of theorems that can already be found in the relevant publications cited in each chapter, we rather create a unified framework for the high-level description of the core idea of each method, and discuss the main requirements and assumptions that need to be satisfied to apply them. This didactic unification of various interval reachability methods was initiated with our recent development of TIRA (*Toolbox for Interval Reachability Analysis*), which is a MATLAB library gathering most of the methods described in this book. The toolbox and its documentation are publicly available at the following address: `https://gitlab.com/pj_meyer/TIRA` and an overview of the toolbox structure and its use is provided in Appendix B.

The field of reachability analysis and the interests in interval reachability analysis are reviewed in Chap. 1, where we also provide the formal definition of the reachability problems considered throughout this book.

Part I describes six main methods for interval reachability analysis. The presentation of all these methods and their respective sub-methods and variations follows the same structure, which focuses on first clearly stating the requirements and limitations for the application of the method, and then listing the main steps needed to obtain the final interval over-approximation of the reachable set. The methods presented in this book rely on interval analysis reviewed in Chap. 2, monotonicity in Chap. 3, mixed monotonicity for continuous-time and discrete-time systems in Chap. 4, and for sampled-data systems in Chap. 5, growth bounds and contraction in Chap. 6, and sampling-based approaches in Chap. 7.

In Part II, we then provide several applications in both academic and more realistic scenarios to illustrate how these reachability methods are used and in what kind of problems they are most relevant. In Chap. 8, reachability analysis is used on both a tunnel diode oscillator and a quadrotor model to verify whether the systems satisfy safety specifications (keeping the state in a safe set at all time) and reachability specifications (reaching a target set in finite time). In Chap. 9, we use the volume of the interval over-approximation of the reachable set to define a measure of the robustness for the controller of a medical exoskeleton. Finally, Chap. 10 presents how reachability analysis plays a central role in the field of abstraction-based control synthesis, where the continuous model of a marine vessel is abstracted into a finite transition system, and a controller for the continuous system with respect to a reach–avoid specification is synthesized by applying model checking algorithms on the finite abstraction.

The intended audience for this book includes graduate students, researchers, and practicing engineers. We believe that the exposition will appeal not only to those with previous experience with reachability analysis or interval methods, but also to those discovering interval reachability analysis for the first time or using it to solve problems similar to those provided in Part II. Previous knowledge and experience of reachability analysis are not required to read this book, since we made our best efforts to present the methods in the most pedagogical manner.

Berkeley, CA, USA   Pierre-Jean Meyer
October 2020   Alex Devonport
                                                                                     Murat Arcak

Acknowledgements

The work of the authors was funded in part by the National Science Foundation grant ECCS-1906164 entitled "Scalable Symbolic Control: Computationally Efficient Design of Feedback Control Algorithms to Satisfy Complex Requirements" (program director: Dr. Kishan Baheti), by the Air Force Office of Scientific Research grant AFOSR FA9550-18-1-0253, entitled "Compositional and Hierarchical Design of Network Control Systems" (program officer: Dr. Frederick A. Leve), and by the Office of Naval Research grant ONR N00014-18-1-2209, entitled "Finite-Horizon Robustness: Moving Beyond Traditional Stability Analysis" (program officer: Dr. Brian Holm-Hansen).

Contents

1   **Introduction**   1
   1.1   Reachability Analysis: Motivation   1
   1.2   Interval Over-Approximation   4
   1.3   Discrete-Time Reachability   6
   1.4   Continuous-Time Reachability   7
   1.5   Definitions and Notation   8
   References   11

Part I   Reachability Methods

2   **Interval Analysis**   15
   2.1   Interval Arithmetics   15
   2.2   Discrete-Time Systems   17
   2.3   Continuous-Time Systems   20
   References   24

3   **Monotonicity**   25
   3.1   Discrete-Time Monotonicity   25
   3.2   Continuous-Time Monotonicity   30
   References   32

4   **Mixed Monotonicity**   33
   4.1   Continuous-Time Mixed Monotonicity   33
   4.2   Discrete-Time Mixed Monotonicity   37
   References   40

5   **Sampled-Data Mixed Monotonicity**   43
   5.1   Sampled-Data Mixed Monotonicity from Bounded
       Sensitivity   43
   5.2   Sensitivity Bounds from Interval Analysis   46

5.3   Sensitivity Bounds from Sampling.....................   48
5.4   Sensitivity Bounds: Hybrid Approach   51
References   58

6   **Growth Bounds**.......................................   61
6.1   Interval Reachability with Growth Bounds   61
6.2   Growth Bounds from Matrix Measures   63
6.3   Growth Bounds from Componentwise Jacobian Bounds   65
6.4   Growth Bounds from Blockwise Matrix Measures   66
References   69

7   **Sampling-Based Methods**   71
7.1   Quasi-Monte Carlo   71
7.2   Monte Carlo....................................   75
References   78

Part II   Applications

8   **Safety and Reachability Verification**   81
8.1   Tunnel Diode Oscillator   82
8.2   Quadrotor Model   84
References   86

9   **Measure of Robustness Against Parameter Uncertainty**   87
9.1   Medical Exoskeleton Model   87
9.2   Robustness Analysis with Sampling-Based Methods   89
References   92

10   **Abstraction-Based Control Synthesis**   93
10.1   Abstraction-Based Control Using Interval Reachability
Analysis..   93
10.2   Reach–Avoid Problem for a Marine Vessel   95
References   100

**Appendix A: Obtaining Sign-Stability by Shifting a Bounded
Variable**...   103

Appendix B: TIRA Toolbox for Interval Reachability Analysis   107

Series Editors' Biographies   111

Chapter 1
Introduction

1.1 Reachability Analysis: Motivation

Reachability analysis deals with the problem of evaluating the set of all the successor states that can be reached in finite time by a system starting from a given set of initial states (Blanchini and Miani 2008). Figure 1.1 illustrates the initial and reachable sets along with possible system evolutions from the initial set to the reachable set.

Reachable sets are of interest in various contexts. For example, the size of the reachable set can be used to evaluate some performance criteria, such as the convergence or divergence of the trajectories, or the robustness of the systems against disturbances or parameter uncertainties (Narvaez-Aroche et al. 2020). Reachable sets are also essential to solve verification or control synthesis problems with respect to safety (ensuring that an unsafe set is never reached) or reachability properties (ensuring that a target set is eventually reached in finite time) (Immler et al. 2018), as well as in the broader field of symbolic control (Tabuada 2009).

Reachable Set Approximations. The exact computation of a reachable set is usually impossible, unless we deal with very specific and simple systems (Lafferriere et al. 2001). Instead, we need to rely on methods to approximate this reachable set. Several directions can be considered when computing such approximations, primarily depending on the objectives and the reasons why we were interested in the evaluation of the reachable set in the first place.

In some cases, the primary objective is to compute a set that approximates the true reachable set as closely as possible. Although methods matching this objective lead to the most accurate approximations of the reachable set, this might come at the cost of a high computational complexity, as well as set representations that may be harder to manipulate (Mitchell et al. 2005). In other cases, the main goal is to reduce the complexity of the reachable set approximation by restricting the approaches to simpler set representations (such as geometrical shapes). The increased ease to

© The Author(s), under exclusive license to Springer Nature Switzerland AG 2021
P.-J. Meyer et al., *Interval Reachability Analysis*,
SpringerBriefs in Control, Automation and Robotics,
https://doi.org/10.1007/978-3-030-65110-7_1

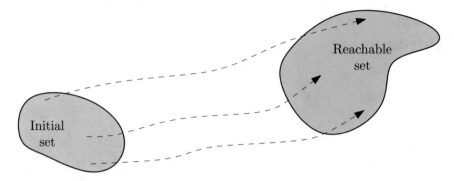

Fig. 1.1 Illustration of the set of successor states that are reachable by a system from the set of initial states. Three system trajectories from the initial set to the reachable set are also represented in dashed lines

manipulate these sets and lower complexity of such approaches come at the cost of a lower accuracy of the approximation.

In this latter category, two subclasses of approaches (over-approximations and under-approximations) predominate in the literature, because elements of analysis on the true reachable set can be deduced from the study of these approximations. In case of an *over-approximation* of the reachable set (see e.g., Girard 2005; Kurzhanskiy and Varaiya 2007), we want the reachable set to be fully contained in the approximation to ensure that we have no *false negatives* (states belonging to the true reachable set but not to its approximation), as illustrated with an ellipse in Fig. 1.2. Over-approximations are particularly well suited to verify the satisfaction of safety objectives, since if the over-approximation does not intersect a set of unsafe

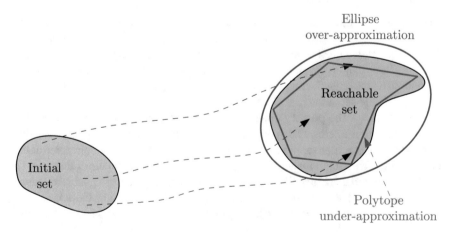

Fig. 1.2 Illustration of an *over-approximation* (here represented as an ellipse) and an *under-approximation* (here represented as a polytope) of the reachable set

states that we want to avoid, then we know that the actual system does not reach any unsafe state either.

Alternatively, we can compute an *under-approximation* (Xue et al. 2017; Goubault and Putot 2017) which is fully included in the true reachable set to ensure that we have no *false positives* (states belonging to the approximation although they are not actually reachable by the system), as illustrated with a polytope in Fig. 1.2. Under-approximations are generally more appropriate to verify that some properties are *not* satisfied: reusing the above safety example, if the under-approximation has a non-empty intersection with the unsafe set, then it is guaranteed that there exists at least one initial state which can reach an unsafe state.

Existing Approaches. The literature on reachability analysis features a great variety of approaches. Some of them consider the problem of evaluating forward reachable sets (states that can be reached from a given initial set through natural evolution of the system) as illustrated in Fig. 1.1 (see e.g., Chen et al. 2012; Althoff 2015), while others focus on the study of reachability problems backward in time, where we are instead given a final set and we want to evaluate all initial states that can reach this set with an appropriate control action (Mitchell et al. 2005; Xue et al. 2017).

The considered systems can be represented either by discrete-time models (difference equations as detailed in Sect. 1.3) (Jaulin et al. 2001; Rakovic et al. 2006), by continuous-time models (differential equations as detailed in Sect. 1.4) (Althoff and Krogh 2011; Shen and Scott 2018), or hybrid models combining both continuous flows and discrete jumps (Stursberg and Krogh 2003; Asarin et al. 2006; Maiga et al. 2015). The reachability problems may either focus on the reachable set at a specific time in the future (or in the past for backward reachability analysis) (Kurzhanskiy and Varaiya 2007; Coogan and Arcak 2015), or on evaluating the *reachable tube* over a whole time range (Althoff et al. 2007; Chen et al. 2012). The problem of evaluating the set of reachable states over an infinite time range is closely related to the literature on invariant sets (Bertsekas 1972; Rakovic et al. 2005).

Numerous set representations are considered to model the approximation of the reachable set. Some of them are based on geometrical shapes for their ease of definition and manipulation, such as hyper-rectangles (Stursberg and Krogh 2003), multidimensional intervals (i.e., axis-aligned hyper-rectangles) as considered in this book, polytopes (Dreossi 2017), zonotopes (Girard 2005) or ellipsoids (Kurzhanskiy and Varaiya 2007). More complex shapes can also be obtained by combining several simpler ones, such as using a paving of intervals of various sizes (Jaulin et al. 2001) or the intersection of zonotopes (Althoff and Krogh 2011). For greater flexibility, other set representations are defined from mathematical functions, such as approaches relying on Taylor models (Chen et al. 2012), support functions (Varaiya 2000) or level sets (Mitchell et al. 2005). The choice of a set representation to use for a reachable set approximation is thus strongly influenced by the desired tradeoff between the accuracy of the approximation and the induced complexity of the corresponding method (in terms of both the computation time and the ease of manipulation of the obtained set).

The reachability analysis problems considered throughout this book focus on the evaluation of **forward** and **finite-time** reachable sets as illustrated in Fig. 1.1. Both **discrete-time** and **continuous-time systems** can be handled by the various methods presented in this book, and most of the considered approaches can deal with fairly general nonlinear dynamics. Finally, the reachability analysis in this book is achieved by computing an **over-approximation** of the reachable set in the form of a multi-dimensional **interval**.

Although the focus of this book is on forward reachability analysis, it is possible to connect such approaches to backward reachability problems. Indeed, as highlighted in Chap. 10, the forward reachability methods considered in this book can be used to abstract a continuous system into a finite transition system on which a backward reachability problem (as well as various other specifications) can be solved.

1.2   Interval Over-Approximation

Using the componentwise inequality $\leq$ and given two vectors $\underline{x}, \overline{x} \in \mathbb{R}^n$ in the n-dimensional Euclidean space $\mathbb{R}^n$ such that $\underline{x} \leq \overline{x}$, the n-dimensional interval

$$\left[\underline{x}, \overline{x}\right] = \left\{x \in \mathbb{R}^n \mid \underline{x} \leq x \leq \overline{x}\right\}$$

is the subset of $\mathbb{R}^n$ containing all the vectors greater than or equal to the lower bound $\underline{x}$ and smaller than or equal to the upper bound $\overline{x}$. Multi-dimensional intervals appear under various names in the literature, such as *axis-aligned hyper-rectangles* or *boxes*.

Representing reachable set over-approximations by intervals has several advantages. Intervals are among the simplest geometrical shapes that can be defined, since their definition only requires the knowledge of two vectors (representing the lower and upper bounds), no matter the dimension of the considered Euclidean space. For numerical implementation, this simplicity also implies that intervals need very low storage and are easy to manipulate. In addition, analyzing the position of any vector with respect to an interval is straightforward as it only relies on component-wise inequalities with the bounds of the interval, and intervals are closed under the intersection operator since the intersection of two intervals is also an interval. The simplicity of this set representation also lowers the complexity of approaches relying on intervals.

On the other hand, the drawbacks of this choice of set representation are its lack of flexibility and the induced conservativeness of the approximations. Indeed, axis-aligned hyper-rectangles are rigid shapes which leave very little freedom. In addition, since our goal is to compute an *over-approximation* of the reachable set (i.e., all reachable states need to belong to its interval approximation), this may result in significant conservativeness due to states that belong to the interval over-approximation despite not actually being reachable. This can, however, be mitigated by splitting the initial set into several smaller subsets and independently computing an interval over-approximation of the reachable set for each of these subsets. A systematic approach relying on this concept of partitioning the state space and applying reachability analysis methods to each partition cell is called *abstraction-based synthesis* and is described in Chap. 10.

Existing Approaches. The literature on interval-based reachability analysis contains a wide range of methods. The main classes of approaches are summarized below. Approaches based on *interval arithmetics* extend basic arithmetic operators (addition, subtraction, multiplication, division) to become applicable to interval variables instead of scalars, and then use these operators to propagate an interval of initial states through the system dynamics (Jaulin et al. 2001; Althoff et al. 2007). In other methods, interval reachability analysis is achieved through the preservation of partial orders by the system description resulting from a *monotonicity property* (Angeli and Sontag 2003) of the dynamics, thus guaranteeing that a reachable set overapproximation can be obtained from the successor states of two extremal states of the initial interval. This approach can be applied to both monotone systems (Moor and Raisch 2002), as well as non-monotone systems, for which an auxiliary monotone system can be created (Coogan and Arcak 2015; Meyer et al. 2018). Some methods focus on the propagation of the initial set over time based on an *upper bound of the growth or contraction* of the distance between trajectories of a continuous-time system (Kapela and Zgliczyński 2009; Maidens and Arcak 2014; Reissig et al. 2016; Fan et al. 2016). Contraction of trajectories toward each other is also known as the property of *incremental stability* (Angeli 2002). Approaches based on *Taylor models* over-approximate reachable sets by computing a finite-order Taylor expansion of the system and then adding a remainder interval bounding the truncation error (Chen et al. 2012). Finally, methods based on *differential inequalities* (Walter 2012) bound the system dynamics by auxiliary differential equations whose solutions define lower and upper bounds of the reachable set for the main system (Scott and Barton 2013; Shen and Scott 2017).

Instead of focusing on reachability analysis, other approaches are primarily designed for validated numerical integration of initial value problems (i.e., solving a differential equation for a specific initial condition while bounding rounding errors due to the numerical integration) and rely, for example, on Taylor models (Nedialkov et al. 1999) or Runge–Kutta integration schemes (dit Sandretto and Chapoutot 2016). While such methods can also be used to solve reachability problems from a set of initial conditions as in this book, this is not their primary intent and they may lead to overly conservative results. In addition, such approaches tend to work best when considering numerical integration over short time ranges, or after splitting a larger time range into smaller sub-steps in order to reduce the accumulated conservativeness over time, known as the *wrapping effect* (Moore et al. 2009). In comparison, this book focuses only on methods for reachability analysis from a set of initial states and where the reachable set over-approximation of continuous-time systems is computed in a single step (i.e., without splitting the time range into smaller steps).

Organization of the Book. The remainder of this introductory chapter provides more formal definitions of the considered reachability problems for discrete-time systems

(Sect. 1.3) and continuous-time systems (Sect. 1.4), and ends with some definitions and notation that are used throughout this book.

Part I of this book presents the theory for six main classes of approaches for interval reachability analysis. Chapter 2 considers the use of interval arithmetics for both discrete-time and continuous-time systems. Chapter 3 focuses on a reachability method relying on the monotonicity property of either discrete-time or continuous-time systems. Chapter 4 is based on a less restrictive property of mixed monotonicity generalizing the results from Chap. 3 to any nonlinear system with bounded Jacobian matrices. Chapter 5 provides an alternative approach to continuous-time mixed monotonicity by instead applying the results for discrete-time mixed monotonicity to the sampled-data version of the continuous-time system, thus converting the bounded Jacobian assumption into requirements on the sensitivity matrix. Chapter 6 considers a method based on a function bounding the growth (or contraction for a negative growth) between trajectories of a continuous-time system. Finally, Chap. 7 focuses on sampling-based reachability methods. For clarity, the methods presented in each of these chapters are described using a unified structure: first stating the main requirements and limitations for the application of the method; next listing the main steps necessary to compute a reachable set over-approximation; and finally providing useful comments and discussions about the results.

Part II provides three concrete scenarios in which interval reachability analysis plays a central role. In Chap. 8, reachability analysis is used to verify whether a system satisfies safety (keeping the state in a safe set at all time) or reachability specifications (reaching a target set in finite time). In Chap. 9, the volume of the interval over-approximation of the reachable set is used to define a measure of the robustness of the controller of a medical exoskeleton. In Chap. 10, we consider an abstraction-based synthesis approach, where reachability analysis is used to represent a continuous system as a finite transition system, and a controller for the continuous system is synthesized by applying model checking algorithms on the finite transition system.

1.3   Discrete-Time Reachability

The first class of systems considered in this book are discrete-time systems defined as a difference equation

$$x^+ = F(t, x, p), \tag{1.1}$$

where the vector field $F : \mathbb{Z} \times \mathbb{R}^{n_x} \times \mathbb{R}^{n_p} \to \mathbb{R}^{n_x}$ describes the successor state vector $x^+ \in \mathbb{R}^{n_x}$ reached by the system starting in initial state $x \in \mathbb{R}^{n_x}$ at time index $t \in \mathbb{Z}$ and with input $p \in \mathbb{R}^{n_p}$. Given an initial time $t_0 \in \mathbb{Z}$ and subsets $X \subseteq \mathbb{R}^{n_x}$ and $P \subseteq \mathbb{R}^{n_p}$, the one-step reachable set of the discrete-time system (1.1) is defined as the set of all successor states reachable from any initial state in X and input in P after a single iteration of the dynamics (1.1),

$$R(t_0, X, P) = \{F(t_0, x, p) \mid x \in X, p \in P\}.$$

Note that this definition refers to the one-step reachable set, for which the final time is $t_0 + 1$. Although the discrete-time reachable set over multiple steps and for a given sequence of inputs can be defined, this definition is not provided here as it is not relevant to the methods presented in the following chapters.

In this book, we focus on interval-based reachability analysis. Therefore, the reachability problem for (1.1) aims to obtain an interval over-approximation of the one-step reachable set when both X and P are defined as intervals of $\mathbb{R}^{n_x}$ and $\mathbb{R}^{n_p}$, respectively.

Problem 1.1 (*Discrete-time reachability*) Given initial time $t_0 \in \mathbb{Z}$, interval of initial states $[\underline{x}, \overline{x}] \subseteq \mathbb{R}^{n_x}$ and interval of input values $[\underline{p}, \overline{p}] \subseteq \mathbb{R}^{n_p}$, find an interval $[\underline{R}, \overline{R}] \subseteq \mathbb{R}^{n_x}$ such that

$$R\left(t_0, [\underline{x}, \overline{x}], [\underline{p}, \overline{p}]\right) \subseteq [\underline{R}, \overline{R}].$$

1.4 Continuous-Time Reachability

We also consider reachability problems for continuous-time systems defined as an ordinary differential equation

$$\dot{x} = f(t, x, p), \tag{1.2}$$

where the vector field $f : \mathbb{R} \times \mathbb{R}^{n_x} \times \mathbb{R}^{n_p} \to \mathbb{R}^{n_x}$ describes the time-derivative of the state trajectories $\dot{x} = \frac{\partial x}{\partial t}$ evaluated at time $t \in \mathbb{R}$, and for state $x \in \mathbb{R}^{n_x}$ and input $p \in \mathbb{R}^{n_p}$. The input vector p can represent either a constant but uncertain parameter of the system or the evaluation of a time-varying signal at time t.

Trajectories of (1.2) are denoted as Φ, where $\Phi(t; t_0, x_0, \mathbf{p})$ represents the state (assumed to exist and be unique) reached at time $t \geq t_0$ by system (1.2) starting from initial state $x_0 \in \mathbb{R}^{n_x}$ at time $t_0 \in \mathbb{R}$ and under piecewise continuous input function $\mathbf{p} : [t_0, +\infty) \to \mathbb{R}^{n_p}$. When the input function $\mathbf{p}$ takes a constant value $p \in \mathbb{R}^{n_p}$ over the time range $[t_0, t]$, we instead write $\Phi(t; t_0, x_0, p)$. Given time range $[t_0, t_f] \subseteq \mathbb{R}$ and subsets $X \subseteq \mathbb{R}^{n_x}$ and $P \subseteq \mathbb{R}^{n_p}$, the reachable set of the continuous-time system (1.2) is defined as the set of all states reachable at time t_f from any initial state in X at time t_0 and any input functions with values in P.

$$R(t_f; t_0, X, P) = \{\Phi(t_f; t_0, x_0, \mathbf{p}) \mid x_0 \in X, \mathbf{p} : [t_0, t_f] \to P\}.$$

Unlike the discrete-time case in Sect. 1.3, the definition of a final time t_f is necessary to specify the end of the time range considered in the continuous-time reachable set.

Similar to the discrete-time case, the reachability problem for (1.2) aims to obtain an interval over-approximation of the reachable set when both X and P are defined as intervals of $\mathbb{R}^{n_x}$ and $\mathbb{R}^{n_p}$, respectively.

Problem 1.2 (*Continuous-time reachability*) Given a time range $[t_0, t_f] \subseteq \mathbb{R}$, interval of initial states $[\underline{x}, \overline{x}] \subseteq \mathbb{R}^{n_x}$ and interval of input values $[\underline{p}, \overline{p}] \subseteq \mathbb{R}^{n_p}$, find an interval $[\underline{R}, \overline{R}] \subseteq \mathbb{R}^{n_x}$ such that

$$R\left(t_f; t_0, [\underline{x}, \overline{x}], [\underline{p}, \overline{p}]\right) \subseteq [\underline{R}, \overline{R}].$$

Although this book is primarily focused on the reachable set at the final instant t_f of the time range, we sometimes also refer to the *reachable tube*

$$\bigcup_{t \in [t_0, t_f]} R\left(t; t_0, [\underline{x}, \overline{x}], [\underline{p}, \overline{p}]\right)$$

corresponding to the union of all reachable sets over the time range $[t_0, t_f]$.

1.5 Definitions and Notation

Ideally, the interval over-approximation of a set $X \subseteq \mathbb{R}^{n_x}$ should be the smallest (in terms of set inclusion) interval containing the set X. We use the word *tight* to describe the over-approximation with this property.

Definition 1.1 (*Tightness*) The interval $[\underline{R}, \overline{R}] \subseteq \mathbb{R}^{n_x}$ is said to be a tight over-approximation of a set $X \subseteq \mathbb{R}^{n_x}$ if for any other interval $[\underline{a}, \overline{a}] \subseteq \mathbb{R}^{n_x}$ it holds that

$$X \subseteq [\underline{a}, \overline{a}] \implies [\underline{R}, \overline{R}] \subseteq [\underline{a}, \overline{a}].$$

This implies that each facet of the hyper-rectangle representing this interval in the space $\mathbb{R}^{n_x}$ has a non-empty intersection with the boundary of X. The tight interval over-approximation of a set is thus uniquely defined.

Figure 1.3 provides an illustration of the introduced concepts for the case of a continuous-time system (1.2) with $n_x = 2$. The elements represented in this figure include the interval $[\underline{x}, \overline{x}]$ of initial states, the corresponding reachable

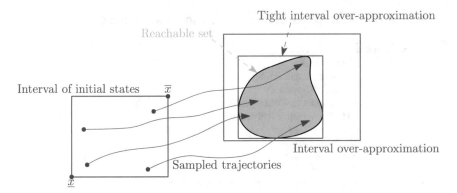

Fig. 1.3 Illustration of the finite-time reachable set from an interval of initial states, and two of its interval over-approximations, one of which is *tight* as in Definition 1.1

set $R\left(t_f; t_0, \left[\underline{x}, \overline{x}\right], \left[\underline{p}, \overline{p}\right]\right)$ (for some input interval $\left[\underline{p}, \overline{p}\right]$, not represented in Fig. 1.3), sampled trajectories of system (1.2) from four initial states picked in $\left[\underline{x}, \overline{x}\right]$ and two interval over-approximations of the reachable set. One of these over-approximations satisfies the tightness property from Definition 1.1 since all four of its facets are in contact with the boundary of the reachable set.

Most of the methods presented in this book to solve Problems 1.1 and 1.2 rely on the study of the Jacobian matrices of systems (1.1) and (1.2), which we define below.

Definition 1.2 Assuming that the vector field F of (1.1) is continuously differentiable, the state Jacobian of the discrete-time system is the matrix representing the partial derivative of the vector field with respect to the state

$$J_x(t, x, p) = \frac{\partial F(t, x, p)}{\partial x} \in \mathbb{R}^{n_x \times n_x}$$

and the input Jacobian is similarly defined to represent variations of the vector field with respect to the input

$$J_p(t, x, p) = \frac{\partial F(t, x, p)}{\partial p} \in \mathbb{R}^{n_x \times n_p}.$$

Since it will always be clear from the context whether we are working with a discrete-time or continuous-time system, the same notations J_x and J_p are used for the Jacobian matrices of a continuous-time system, simply replacing F by the vector field f of (1.2) in the above definitions.

We end this chapter with a list of the main notation used throughout this book.

Common Sets and Set Operations

$\mathbb{Z}$	Set of integers
$\mathbb{N}$	Set of nonnegative integers
$\mathbb{R}$	Set of real numbers
$\mathbb{R}_+$	Set of nonnegative real numbers: $\mathbb{R}_+ = \{x \in \mathbb{R} \mid x \geq 0\}$
$\mathbb{R}_-$	Set of nonpositive real numbers: $\mathbb{R}_- = \{x \in \mathbb{R} \mid x \leq 0\}$
$\emptyset$	Empty set
$\subseteq$	Set inclusion
2^X	Set of subsets of X: $2^X = \{Y \subseteq X\}$
$\cup$	Set union: $X \cup Y = \{z \mid z \in X \text{ or } z \in Y\}$
$\cap$	Set intersection: $X \cap Y = \{z \mid z \in X \text{ and } z \in Y\}$
$+$	Set addition: $X + Y = \{x + y \mid x \in X, y \in Y\}$
$\backslash$	Set difference: $X \backslash Y = \{x \in X \mid x \notin Y\}$
$\times$	Cartesian product: $X \times Y = \{(x, y) \mid x \in X, y \in Y\}$

Vector and Matrix Spaces

$\mathbb{R}^n$	Set of n-dimensional real vectors
$\mathbb{R}^{m \times n}$	Set of $m \times n$ real matrices
$\mathbf{0}_n$	n-dimensional vector filled with zeros
$\mathbf{1}_n$	n-dimensional vector filled with ones
I_n	Identity matrix in $\mathbb{R}^{n \times n}$
a_i	ith component of vector a
A_{ij}	Component (i, j) of matrix A
$[a_1, \ldots, a_n]$	Inline notation of a row vector in $\mathbb{R}^{1 \times n}$
$[a_1; \ldots; a_n]$	Inline notation of a column vector in $\mathbb{R}^n$

Operations in Vector and Matrix Spaces

$\geq, \leq, >, <$	Componentwise inequalities of vectors or matrices
$\preceq_x, \preceq_p$	Partial orders on vector spaces, used for monotonicity in Chap. 3
$\otimes$	Kronecker product of two matrices
$\lvert a \rvert$	Componentwise absolute value of vector or matrix a
$\lVert a \rVert_\infty$	Infinity norm of vector a: $\lVert a \rVert_\infty = \max_i \lvert a_i \rvert$
$\lVert A \rVert_\infty$	Infinity norm of matrix A: $\lVert A \rVert_\infty = \max_i \sum_j \lvert A_{ij} \rvert$
$\lVert [\underline{A}, \overline{A}] \rVert_\infty$	Infinity norm of interval matrix $[\underline{A}, \overline{A}]$: $\lVert [\underline{A}, \overline{A}] \rVert_\infty = \lVert \max (\lvert \underline{A} \rvert, \lvert \overline{A} \rvert) \rVert_\infty$
$\mu_\infty(A)$	Infinity matrix measure of matrix A: $\mu_\infty(A) = \max_i \left(A_{ii} + \sum_{i \neq j} \lvert A_{ij} \rvert \right)$

Intervals

$[\underline{x}, \overline{x}]$	Scalar, vector or matrix interval: $[\underline{x}, \overline{x}] = \{x \mid \underline{x} \leq x \leq \overline{x}\}$
$\mathbb{I}$	Set of intervals in $\mathbb{R}$
x^*	Center of the interval $[\underline{x}, \overline{x}]$: $x^* = \frac{\underline{x}+\overline{x}}{2}$
$[x]$	Half-width of the interval $[\underline{x}, \overline{x}]$: $[x] = \frac{\overline{x}-\underline{x}}{2}$
H	Interval hull of one or several sets

System Description

t, t_0, t_f, τ	Time in $\mathbb{R}$ for continuous-time systems, in $\mathbb{Z}$ for discrete-time systems
$x, x_0, \underline{x}, \overline{x}, \xi^i$	State of the system, in $\mathbb{R}^{n_x}$
$p, \underline{p}, \overline{p}, \pi^i$	Input of the system, in $\mathbb{R}^{n_p}$
$\mathbf{p}$	Input signal $\mathbf{p} : \mathbb{R} \to \mathbb{R}^{n_p}$ for continuous-time systems
x^+	One-step successor of state x for a discrete-time system
$\dot{x}$	Time derivative of state x for a continuous-time system
F	Vector field of a discrete-time system
f	Vector field of a continuous-time system
Φ	State trajectory of a continuous-time system

R	Reachable set of discrete-time or continuous-time system
J_x, J_p	Jacobian matrices with respect to state and input
J_{xx}	Second-order Jacobian matrix (Sect. 5.4)
S_x	Sensitivity matrix with respect to initial states (Chap. 5)
S_{xx}	Second-order sensitivity matrix (Sect. 5.4)
L_x, L_p	Shifting matrices for mixed monotonicity (Chaps. 4–5)
g	Mixed-monotonicity decomposition function (Chaps. 4–5)
G	Growth bound function (Chap. 6)

References

Althoff M (2015) An introduction to CORA 2015. In: ARCH@ CPSWeek, pp 120–151

Althoff M, Krogh BH (2011) Zonotope bundles for the efficient computation of reachable sets. In: 2011 50th IEEE conference on decision and control and European control conference. IEEE, pp 6814–6821

Althoff M, Stursberg O, Buss M (2007) Reachability analysis of linear systems with uncertain parameters and inputs. In: 46th IEEE conference on decision and control. IEEE, pp 726–732

Angeli D (2002) A Lyapunov approach to incremental stability properties. IEEE Trans Autom Control 47(3):410–421

Angeli D, Sontag ED (2003) Monotone control systems. IEEE Trans Autom Control 48(10):1684–1698

Asarin E, Dang T, Frehse G, Girard A, Le Guernic C, Maler O (2006) Recent progress in continuous and hybrid reachability analysis. In: 2006 IEEE conference on computer aided control system design, 2006 ieee international conference on control applications, 2006 ieee international symposium on intelligent control. IEEE, pp 1582–1587

Bertsekas D (1972) Infinite time reachability of state-space regions by using feedback control. IEEE Trans Autom Control 17(5):604–613

Blanchini F, Miani S (2008) Set-theoretic methods in control. Springer

Chen X, Abraham E, Sankaranarayanan S (2012) Taylor model flowpipe construction for non-linear hybrid systems. In: 2012 IEEE 33rd real-time systems symposium. IEEE, pp 183–192

Coogan S, Arcak M (2015) Efficient finite abstraction of mixed monotone systems. In: 18th international conference on hybrid systems: computation and control, pp 58–67

dit Sandretto JA, Chapoutot A (2016) Validated explicit and implicit Runge-Kutta methods. Reliab Comput 22(1):79–103

Dreossi T (2017) Sapo: reachability computation and parameter synthesis of polynomial dynamical systems. In: 20th international conference on hybrid systems: computation and control. ACM, pp 29–34

Fan C, Kapinski J, Jin X, Mitra S (2016) Locally optimal reach set over-approximation for nonlinear systems. In: 2016 international conference on embedded software (EMSOFT). IEEE, pp 1–10

Girard A (2005) Reachability of uncertain linear systems using zonotopes. In: International workshop on hybrid systems: computation and control. Springer, pp 291–305

Goubault E, Putot S (2017) Forward inner-approximated reachability of non-linear continuous systems. In: Proceedings of the 20th international conference on hybrid systems: computation and control, pp 1–10

Immler F, Althoff M, Chen X, Fan C, Frehse G, Kochdumper N, Li Y, Mitra S, Tomar MS, Zamani M (2018) ARCH-COMP18 category report: continuous and hybrid systems with nonlinear dynamics. In: Proceedings of the 5th international workshop on applied verification for continuous and hybrid systems

Jaulin L, Kieffer M, Didrit O, Walter E (2001) Applied interval analysis: with examples in parameter and state estimation, robust control and robotics, vol 1. Springer Science & Business Media

Kapela T, Zgliczyński P (2009) A Lohner-type algorithm for control systems and ordinary differential inclusions. Discrete Contin Dyn Syst Ser B 11(2):365–385

Kurzhanskiy AA, Varaiya P (2007) Ellipsoidal techniques for reachability analysis of discrete-time linear systems. IEEE Trans Autom Control 52(1):26–38

Lafferriere G, Pappas GJ, Yovine S (2001) Symbolic reachability computation for families of linear vector fields. J Symb Comput 32(3):231–253

Maidens J, Arcak M (2014) Reachability analysis of nonlinear systems using matrix measures. IEEE Trans Autom Control 60(1):265–270

Maiga M, Ramdani N, Travé-Massuyès L, Combastel C (2015) A comprehensive method for reachability analysis of uncertain nonlinear hybrid systems. IEEE Trans Autom Control 61(9):2341–2356

Meyer PJ, Coogan S, Arcak M (2018) Sampled-data reachability analysis using sensitivity and mixed-monotonicity. IEEE Control Syst Lett 2(4):761–766

Mitchell IM, Bayen AM, Tomlin CJ (2005) A time-dependent Hamilton-Jacobi formulation of reachable sets for continuous dynamic games. IEEE Trans Autom Control 50(7):947–957

Moor T, Raisch J (2002) Abstraction based supervisory controller synthesis for high order monotone continuous systems. In: Modelling, analysis, and design of hybrid systems. Springer, pp 247–265

Moore RE, Kearfott RB, Cloud MJ (2009) Introduction to interval analysis, vol 110. Siam

Narvaez-Aroche O, Meyer PJ, Tu S, Packard A, Arcak M (2020) Robust control of the sit-to-stand movement for a powered lower limb orthosis. IEEE Trans Control Syst Technol 28(6):2390–2403

Nedialkov NS, Jackson KR, Corliss GF (1999) Validated solutions of initial value problems for ordinary differential equations. Appl Math Comput 105(1):21–68

Rakovic SV, Kerrigan EC, Kouramas KI, Mayne DQ (2005) Invariant approximations of the minimal robust positively invariant set. IEEE Trans Autom Control 50(3):406–410

Rakovic SV, Kerrigan EC, Mayne DQ, Lygeros J (2006) Reachability analysis of discrete-time systems with disturbances. IEEE Trans Autom Control 51(4):546–561

Reissig G, Weber A, Rungger M (2016) Feedback refinement relations for the synthesis of symbolic controllers. IEEE Trans Autom Control 62(4):1781–1796

Scott JK, Barton PI (2013) Bounds on the reachable sets of nonlinear control systems. Automatica 49(1):93–100

Shen K, Scott JK (2017) Rapid and accurate reachability analysis for nonlinear dynamic systems by exploiting model redundancy. Comput Chem Eng 106:596–608

Shen K, Scott JK (2018) Mean value form enclosures for nonlinear reachability analysis. In: 2018 IEEE conference on decision and control (CDC). IEEE, pp 7112–7117

Stursberg O, Krogh BH (2003) Efficient representation and computation of reachable sets for hybrid systems. In: International workshop on hybrid systems: computation and control. Springer, pp 482–497

Tabuada P (2009) Verification and control of hybrid systems: a symbolic approach. Springer Science & Business Media

Varaiya P (2000) Reach set computation using optimal control. In: Verification of digital and hybrid systems. Springer, pp 323–331

Walter W (2012) Differential and integral inequalities, vol 55. Springer Science & Business Media

Xue B, She Z, Easwaran A (2017) Underapproximating backward reachable sets by semialgebraic sets. IEEE Trans Autom Control 62(10):5185–5197

Part I
Reachability Methods

Chapter 2
Interval Analysis

This chapter presents reachability analysis methods to over-approximate reachable sets by intervals using classical notions from the field of *interval analysis*. Unlike other reachability analysis methods presented in later chapters, which rely on mathematical properties satisfied by the system for the computation of these over-approximations, interval analysis is a more straightforward approach where state and input variables bounded in intervals are propagated through the system dynamics to obtain an interval over-approximation of the reachable set. This propagation is obtained by extending the definitions of the four arithmetic operators (addition, subtraction, multiplication, division) to be applicable to interval variables, or more generally to interval matrices.

Although the reachability approaches presented in this chapter can be used on their own, the resulting over-approximations are often too conservative for their practical use. This chapter thus primarily aims to introduce these methods as tools to be used within other reachability methods presented later on in this book.

The main arithmetic operations for interval matrices are introduced in Sect. 2.1. Two methods for the over-approximation of reachable sets of discrete-time and continuous-time systems using interval analysis are then presented in Sects. 2.2 and 2.3, respectively.

2.1 Interval Arithmetics

We start by providing the definitions for various arithmetic operations on intervals and interval matrices. We first recall the definition of a (closed) scalar interval as the set $[\underline{a}, \overline{a}] = \{a \in \mathbb{R} \mid \underline{a} \leq a \leq \overline{a}\}$. An interval matrix is identically defined using componentwise inequalities $[\underline{A}, \overline{A}] = \{A \in \mathbb{R}^{m \times n} \mid \underline{A} \leq A \leq \overline{A}\}$. We denote

P.-J. Meyer et al., *Interval Reachability Analysis*,
SpringerBriefs in Control, Automation and Robotics,
https://doi.org/10.1007/978-3-030-65110-7_2

as $\mathbb{I} \subseteq 2^{\mathbb{R}}$ the set of all closed scalar intervals in $\mathbb{R}$ and $\mathbb{I}^{m \times n}$ the set of all interval matrices in $\mathbb{R}^{m \times n}$.

The product of a scalar $a \in \mathbb{R}$ and an interval matrix $\left[\underline{A}, \overline{A}\right] \in \mathbb{I}^{m \times n}$ is the interval matrix in $\mathbb{I}^{m \times n}$

$$a * \left[\underline{A}, \overline{A}\right] = \begin{cases} \left[a * \underline{A}, a * \overline{A}\right] & \text{if } a \geq 0, \\ \left[a * \overline{A}, a * \underline{A}\right] & \text{if } a \leq 0. \end{cases}$$

The sum of two interval matrices $\left[\underline{A}, \overline{A}\right], \left[\underline{B}, \overline{B}\right] \in \mathbb{I}^{m \times n}$ of the same size is the interval matrix in $\mathbb{I}^{m \times n}$

$$\left[\underline{A}, \overline{A}\right] + \left[\underline{B}, \overline{B}\right] = \left[\underline{A} + \underline{B}, \overline{A} + \overline{B}\right].$$

Subtracting two interval matrices $\left[\underline{A}, \overline{A}\right], \left[\underline{B}, \overline{B}\right] \in \mathbb{I}^{m \times n}$ of the same size is obtained by combining the above two operations ($\left[\underline{A}, \overline{A}\right] - \left[\underline{B}, \overline{B}\right] = \left[\underline{A}, \overline{A}\right] + (-1) * \left[\underline{B}, \overline{B}\right]$), which results in

$$\left[\underline{A}, \overline{A}\right] - \left[\underline{B}, \overline{B}\right] = \left[\underline{A} - \overline{B}, \overline{A} - \underline{B}\right].$$

The product of two scalar intervals $\left[\underline{a}, \overline{a}\right], \left[\underline{b}, \overline{b}\right] \in \mathbb{I}$ is defined as

$$\left[\underline{a}, \overline{a}\right] * \left[\underline{b}, \overline{b}\right] = \left[\min\left(\underline{ab}, \underline{a}\overline{b}, \overline{a}\underline{b}, \overline{ab}\right), \max\left(\underline{ab}, \underline{a}\overline{b}, \overline{a}\underline{b}, \overline{ab}\right)\right].$$

This scalar interval product is then used to define the product of two interval matrices $\left[\underline{A}, \overline{A}\right] \in \mathbb{I}^{m \times n}$ and $\left[\underline{B}, \overline{B}\right] \in \mathbb{I}^{n \times p}$ as an interval matrix in $\mathbb{I}^{m \times p}$ defined componentwise

$$\left(\left[\underline{A}, \overline{A}\right] * \left[\underline{B}, \overline{B}\right]\right)_{ij} = \sum_{k=1}^{n} \left[\underline{A}_{ik}, \overline{A}_{ik}\right] * \left[\underline{B}_{kj}, \overline{B}_{kj}\right].$$

The product of a scalar interval $\left[\underline{a}, \overline{a}\right] \in \mathbb{I}$ with an interval matrix $\left[\underline{A}, \overline{A}\right] \in \mathbb{I}^{m \times n}$ is the interval matrix in $\mathbb{I}^{m \times n}$ defined componentwise

$$\left(\left[\underline{a}, \overline{a}\right] * \left[\underline{A}, \overline{A}\right]\right)_{ij} = \left[\underline{a}, \overline{a}\right] * \left[\underline{A}_{ij}, \overline{A}_{ij}\right].$$

The power operator of a square interval matrix $\left[\underline{A}, \overline{A}\right] \in \mathbb{I}^{m \times m}$ can be reduced to a finite number of interval matrix products,

$$\left[\underline{A}, \overline{A}\right]^{i} = \left[\underline{A}, \overline{A}\right]^{i-1} * \left[\underline{A}, \overline{A}\right] \quad \text{or} \quad \left[\underline{A}, \overline{A}\right]^{i} = \left[\underline{A}, \overline{A}\right] * \left[\underline{A}, \overline{A}\right]^{i-1}.$$

While these two expressions may return different results, both over-approximate the desired set $\left\{A^{i} \mid A \in \left[\underline{A}, \overline{A}\right]\right\}$. Therefore, whenever a power interval matrix $\left[\underline{A}, \overline{A}\right]^{i}$ appears in this book, we do not specify which of these two definitions is used, since both are acceptable.

The inverse of a scalar interval $[\underline{a}, \overline{a}] \in \mathbb{I}$ is the scalar interval (possibly empty or unbounded)

$$1/[\underline{a}, \overline{a}] = \begin{cases} [1/\overline{a}, 1/\underline{a}] & \text{if } 0 \notin [\underline{a}, \overline{a}], \\ \emptyset & \text{if } [\underline{a}, \overline{a}] = [0, 0], \\ [1/\overline{a}, +\infty) & \text{if } \underline{a} = 0 \text{ and } \overline{a} > 0, \\ (-\infty, 1/\underline{a}] & \text{if } \underline{a} < 0 \text{ and } \overline{a} = 0, \\ (-\infty, +\infty) & \text{if } \underline{a} < 0 < \overline{a}. \end{cases}$$

The inverse of an interval matrix is defined componentwise. Given two interval matrices $[\underline{A}, \overline{A}] \in \mathbb{I}^{m \times n}$ and $[\underline{B}, \overline{B}] \in \mathbb{I}^{n \times p}$, the division of $[\underline{A}, \overline{A}]$ by $[\underline{B}, \overline{B}]$ is the interval matrix in $\mathbb{I}^{m \times p}$ obtained by first inverting $[\underline{B}, \overline{B}]$, and then taking the product with $[\underline{A}, \overline{A}]$,

$$[\underline{A}, \overline{A}] / [\underline{B}, \overline{B}] = [\underline{A}, \overline{A}] * \left(1 / [\underline{B}, \overline{B}]\right).$$

Given two interval matrices $[\underline{A}, \overline{A}] \in \mathbb{I}^{m \times n}$ and $[\underline{B}, \overline{B}] \in \mathbb{I}^{p \times q}$, their interval Kronecker product is an interval matrix in $\mathbb{I}^{mp \times nq}$ defined as an $m \times n$ block interval matrix whose (i, j) block is in $\mathbb{I}^{p \times q}$ and defined as

$$\left([\underline{A}, \overline{A}] \otimes [\underline{B}, \overline{B}]\right)_{ij} = [\underline{A}_{ij}, \overline{A}_{ij}] * [\underline{B}, \overline{B}].$$

Finally, the interval hull (also referred to as *union* in Jaulin et al. 2001) of two interval matrices $[\underline{A}, \overline{A}], [\underline{B}, \overline{B}] \in \mathbb{I}^{m \times n}$ of the same size is introduced as the operator $H : \mathbb{I}^{m \times n} \times \mathbb{I}^{m \times n} \to \mathbb{I}^{m \times n}$ defined as

$$H\left([\underline{A}, \overline{A}], [\underline{B}, \overline{B}]\right) = \left[\min\left(\underline{A}, \underline{B}\right), \max\left(\overline{A}, \overline{B}\right)\right],$$

using componentwise min and max operators.

2.2 Discrete-Time Systems

Using the arithmetic operations defined in Sect. 2.1, the one-step reachable set of the discrete-time system (1.1) can be directly expressed by replacing the initial state and input by their interval bounds in the vector field,

$$R\left(t_0, [\underline{x}, \overline{x}], [\underline{p}, \overline{p}]\right) = \left\{F\left(t_0, x, p\right) \mid x \in [\underline{x}, \overline{x}], p \in [\underline{p}, \overline{p}]\right\}$$
$$= F\left(t_0, [\underline{x}, \overline{x}], [\underline{p}, \overline{p}]\right).$$

The main difficulty arising from this approach is when the vector field of the discrete-time system cannot be defined solely with the four arithmetic operators, but also relies on other elementary functions such as exponentials, square roots or trigonometric functions.

Requirements and Limitations. If the vector field of (1.1) can be written as an equation involving only a finite number of the four arithmetic operators (addition, subtraction, multiplication, division) between constant and variables matrices, then the method presented below has no further requirement.

On the other hand, if the vector field can only be written through the use of external functions which cannot be decomposed into finitely many arithmetic operations, then the user is required to provide methods for the over-approximation of these functions. Among the most classical examples of such functions that frequently appear in system dynamics are continuous and monotone functions (exponential, square root), continuous but non-monotone functions (trigonometric functions), continuous but non-smooth functions (absolute value), and discontinuous functions (sign).

For a scalar function $g : \mathbb{R} \to \mathbb{R}$, the user thus needs to provide an inclusion function $[g] : \mathbb{I} \to \mathbb{I}$ such that for any interval $\left[\underline{a}, \overline{a}\right] \subseteq \mathbb{R}$, $[g]\left(\left[\underline{a}, \overline{a}\right]\right)$ is an interval over-approximation of the image of g,

$$\left\{ g(a) \mid a \in \left[\underline{a}, \overline{a}\right] \right\} \subseteq [g]\left(\left[\underline{a}, \overline{a}\right]\right).$$

For monotone functions (whether continuous as the exponential and square root, or discontinuous as the sign function), such inclusion functions can be trivially defined as $[g]\left(\left[\underline{a}, \overline{a}\right]\right) = \left[g\left(\underline{a}\right), g\left(\overline{a}\right)\right]$ if g is increasing and $[g]\left(\left[\underline{a}, \overline{a}\right]\right) = \left[g\left(\overline{a}\right), g\left(\underline{a}\right)\right]$ if g is decreasing. For non-monotone functions or non-scalar functions ($g : \mathbb{R}^m \to \mathbb{R}^n$), the definition of these inclusion functions might not be as straightforward and specific algorithms need to be provided by the user. If we consider the example of the cosine function, we can define $[\cos]\left(\left[\underline{a}, \overline{a}\right]\right) = \left[\underline{c}, \overline{c}\right]$ with

$$\underline{c} = \begin{cases} -1 \text{ if } \exists k \in \mathbb{Z} \mid 2k\pi + \pi \in \left[\underline{a}, \overline{a}\right], \\ \min\left(\cos\left(\underline{a}\right), \cos\left(\overline{a}\right)\right) \text{ otherwise,} \end{cases} \quad \overline{c} = \begin{cases} 1 \text{ if } \exists k \in \mathbb{Z} \mid 2k\pi \in \left[\underline{a}, \overline{a}\right], \\ \max\left(\cos\left(\underline{a}\right), \cos\left(\overline{a}\right)\right) \text{ otherwise.} \end{cases}$$

Reachability Method. Once inclusion functions for any elementary function appearing in the vector field of the system have been provided, replacing these functions in F by their inclusion functions results in the definition of an inclusion function for the vector field. This inclusion function of F can thus be used as in Proposition 2.1 below to over-approximate the reachable set of (1.1) for any pair of state and input intervals as defined in Problem 1.1.

Proposition 2.1 *Assume that the discrete-time system (1.1) is defined as a finite composition of the four arithmetic operators (addition, subtraction, multiplication, division) and elementary functions for which inclusion functions are provided. In the expression of $F(t_0, x, p)$, replace x and p by their respec-*

tive intervals $[\underline{x}, \overline{x}]$ and $[\underline{p}, \overline{p}]$ from Problem 1.1 and replace the elementary functions by their inclusion functions. Then the computation of the resulting expression using the interval arithmetic operators is an interval over-approximation of the one-step reachable set $R\left(t_0, [\underline{x}, \overline{x}], [\underline{p}, \overline{p}]\right)$ of (1.1) solving Problem 1.1.

Discussion. Although this method is applicable to very general discrete-time systems as long as inclusion functions for their elementary functions are provided, it has the drawback of sometimes resulting in very conservative over-approximations. This conservativeness is primarily influenced by the number of occurrences of each variable in the vector field. In particular, the operator for interval multiplication is not distributive and we have

$$[\underline{a}, \overline{a}] * ([\underline{b}, \overline{b}] + [\underline{c}, \overline{c}]) \subseteq [\underline{a}, \overline{a}] * [\underline{b}, \overline{b}] + [\underline{a}, \overline{a}] * [\underline{c}, \overline{c}].$$

This means that, one way to reduce the conservativeness of the result is by factorizing the expression of the vector field as much as possible to minimize the number of occurrences of each variable.

More generally, two algebraic expressions which are equal when their variables are real may lose their equality if the variables are replaced by intervals. For example, we know that $(a + 1)^2 = a^2 + 2a + 1$ for all $a \in \mathbb{R}$, but if we replace a by the interval $[-1, 1]$, then $([-1, 1] + 1)^2 = [0, 4]$, while $[-1, 1]^2 + 2 * [-1, 1] + 1 = [-2, 4]$. However, both results from the interval expressions are guaranteed to contain the set $\{(a + 1)^2 \mid a \in [-1, 1]\}$. Therefore, in cases where the factorization of the vector field is too difficult, a second option to reduce the conservativeness is to compute the interval over-approximation as in Proposition 2.1 for several redundant expressions of the vector field. Then, the intersection of all the resulting intervals is still a solution to Problem 1.1.

A third option to reduce conservativeness of the results is to use the mean-value theorem to define an inclusion function of the vector field, expressed with its first-order derivatives (the Jacobian matrices of the system) and the evaluation of the vector field at the center of both intervals $F(t_0, x^*, p^*)$. While this approach can reduce the conservativeness, especially when the widths of intervals $[\underline{x}, \overline{x}]$ and $[\underline{p}, \overline{p}]$ are small, it comes with two additional limitations. The first one is that the vector field needs to be differentiable. The second drawback is that this method relies on the knowledge of an inclusion function for the derivatives of F, which needs to be provided by the user. Compared to Proposition 2.1, this approach based on the mean-value theorem thus tends to simply shift the problem of computing an inclusion function for F as in Proposition 2.1, to computing inclusion functions of the Jacobian matrices J_x and J_p.

A generalization of this approach is obtained by iterating this mean-value theorem method to successively express inclusion functions of the ith-order derivative based on one of the $(i + 1)$th-order derivatives. This results in an inclusion function of F defined as a Taylor series of some order r which incorporates evaluations of F and its first- to $(r - 1)$th-order derivatives at the center of intervals $[\underline{x}, \overline{x}]$ and $[\underline{p}, \overline{p}]$, as well as an inclusion function of the rth-order derivative. Again, this generalization requires the vector field to be at least r times differentiable and for the user to be able to provide an inclusion function of its rth-order derivative. More details on these alternative approaches can be found in Jaulin et al. (2001).

2.3 Continuous-Time Systems

For continuous-time systems, the use of interval arithmetics for reachability analysis is not as straightforward as in the discrete-time case. Indeed, the differential equation (1.2) describes the time derivative of the state, and applying the interval arithmetic operator to this vector field as in Sect. 2.2 would not provide any useful knowledge about the reachable set. In addition, a closed-form expression of the trajectory function Φ is usually not available for most continuous-time systems, which implies that using interval arithmetics on Φ is also not an option.

For these reasons and despite the apparent generality of the discrete-time results in Sect. 2.2, this section on continuous-time systems only focuses on the restricted class of affine systems. This particular case is admissible because a closed-form expression of Φ is known for such systems.

Requirements and Limitations. This method is applicable to any continuous-time system whose vector field is affine in the state

$$\dot{x} = A(t)x + B(t), \tag{2.1}$$

whose parameters $A(t) \in \mathbb{R}^{n_x \times n_x}$ and $B(t) \in \mathbb{R}^{n_x}$ may be time-varying. Although this affine system does not explicitly have input variables as in the general continuous-time system (1.2), the time-varying matrix $A(t)$ and vector $B(t)$ can be seen as inputs of (2.1). Lastly, we need to make the assumption that $A(t)$ and $B(t)$ are bounded.

Assumption 2.1 There exists intervals $[\underline{A}, \overline{A}] \subseteq \mathbb{R}^{n_x \times n_x}$ and $[\underline{B}, \overline{B}] \subseteq \mathbb{R}^{n_x}$ bounding the time-varying matrices $A(t) \in \mathbb{R}^{n_x \times n_x}$ and $B(t) \in \mathbb{R}^{n_x}$ at all time,

$$\forall t \in \mathbb{R}, \ A(t) \in [\underline{A}, \overline{A}], \ B(t) \in [\underline{B}, \overline{B}].$$

To simplify the notation in this section, we introduce shorthand notation for the following three intervals: $\mathcal{X}_0 = [\underline{x}, \overline{x}]$ from Problem 1.2, $\mathcal{A} = [\underline{A}, \overline{A}]$ and $\mathcal{B} = [\underline{B}, \overline{B}]$ from Assumption 2.1.

Reachability Method. The main idea of this approach is to consider the closed-form expression of the solutions of (2.1) and then replace $A(t)$, $B(t)$, and x_0 by their respective intervals $\mathscr{A}$, $\mathscr{B}$, and $\mathscr{X}_0$ to finally compute an interval over-approximation of the reachable set using the interval arithmetic operators from Sect. 2.1. Since such solution of (2.1) contains matrix exponentials, a preliminary step is to over-approximate their interval expression using a truncated Taylor series expansion in order to obtain an expression which contains only a finite number of interval arithmetic operators. The results below hold if the truncation of these Taylor series is done at an order $r \in \mathbb{N}$ satisfying $r > \|\mathscr{A}\|_\infty (t_f - t_0) - 2$, where the infinity norm of the interval matrix is defined as $\|\mathscr{A}\|_\infty = \left\| \max \left(|\underline{A}|, |\overline{A}| \right) \right\|_\infty$ using componentwise functions for the absolute value and maximum.

We first introduce three functions $C_1, C_2, C_3 : \mathbb{R}_+ \to \mathbb{I}^{n_x \times n_x}$ related to these Taylor expansions

$$C_1(\tau) = \left[-\mathbf{1}_{n_x \times n_x}, \mathbf{1}_{n_x \times n_x} \right] * \frac{(\|\mathscr{A}\|_\infty \tau)^{r+1}}{(r+1)!} \frac{r+2}{r+2 - \|\mathscr{A}\|_\infty \tau},$$

$$C_2(\tau) = \sum_{i=0}^{r} \frac{(\mathscr{A}\tau)^i}{i!} + C_1(\tau),$$

$$C_3(\tau) = \sum_{i=0}^{r} \frac{\mathscr{A}^i \tau^{i+1}}{(i+1)!} + C_1(\tau)\tau,$$

which take a positive time range ($\tau = t_f - t_0$ in the case of Problem 1.2) as input argument and return an interval in $\mathbb{R}^{n_x \times n_x}$. These functions are then used to express an over-approximation of the reachable set of (2.1).

Proposition 2.2 *Under Assumption 2.1 and with Taylor order $r > \|\mathscr{A}\|_\infty (t_f - t_0) - 2$, the interval*

$$C_2(t_f - t_0) * \mathscr{X}_0 + C_3(t_f - t_0) * \mathscr{B}$$

is an over-approximation of the reachable set $R(t_f; t_0, \mathscr{X}_0)$ of the continuous-time affine system (2.1) solving Problem 1.2.

An interpretation of this result can be obtained by comparing it with the definition of successors of (2.1) when intervals $\mathscr{A}$, $\mathscr{B}$, and $\mathscr{X}_0$ are reduced to singletons. If we take $\mathscr{A} = \{A\}$, $\mathscr{B} = \{B\}$, and $\mathscr{X}_0 = \{x_0\}$, then the successor of (2.1) with constant matrices A and B and initial state x_0 is

$$\Phi(t_f; t_0, x_0) = e^{A*(t_f - t_0)} x_0 + \int_{t_0}^{t_f} e^{A*t} B \, dt.$$

The result in Proposition 2.2 can thus be seen as the generalization of this result by replacing A, B, and x_0 by their respective intervals $\mathscr{A}$, $\mathscr{B}$, and $\mathscr{X}_0$,

$$R\left(t_f; t_0, \mathscr{X}_0\right) \subseteq e^{\mathscr{A}*(t_f-t_0)} * \mathscr{X}_0 + \int_{t_0}^{t_f} e^{\mathscr{A}*t} * \mathscr{B}dt.$$

Indeed, the term $C_2(t_f - t_0)$ is an over-approximation of the interval matrix exponential $e^{\mathscr{A}*(t_f-t_0)}$ composed of two terms: the sum is the Taylor series expansion of the exponential truncated at order r; and the second term $C_1(t_f - t_0)$ is an over-approximation of the remaining terms of the Taylor series after the truncation. Finally, the term $C_3(t_f - t_0)$ is an over-approximation of the integral of $C_2(\tau)$ for $\tau \in [t_0, t_f]$.

Although the remainder of the book primarily focuses on the reachable set at the final time t_f, the method in this section also allows an over-approximation of the reachable tube over the whole time range $[t_0, t_f]$. This reachable tube is defined as the union of all the reachable sets from the initial time t_0 to the final time t_f,

$$\bigcup_{t \in [t_0, t_f]} R\left(t; t_0, \mathscr{X}_0\right).$$

In addition to functions C_1, C_2 and C_3 introduced for the reachable set over-approximation in Proposition 2.2, the computation of an over-approximation of this reachable tube relies on the two functions $C_4, C_5 : \mathbb{R}_+ \to \mathbb{I}^{n_x \times n_x}$

$$C_4(\tau) = \left[\sum_{i=2}^{r} \left(i^{\frac{-i}{i-1}} - i^{\frac{-1}{i-1}}\right) \frac{(\mathscr{A}\tau)^i}{i!}, \mathbf{0}_{n_x \times n_x}\right] + C_1(\tau),$$

$$C_5(\tau) = \left[\sum_{i=2}^{r} \left(i^{\frac{-i}{i-1}} - i^{\frac{-1}{i-1}}\right) \frac{\mathscr{A}^{i-1}\tau^i}{i!}, \mathbf{0}_{n_x \times n_x}\right] + \frac{C_1(\tau)}{\|\mathscr{A}\|_\infty}.$$

Proposition 2.3 *Under Assumption 2.1, define* $\tilde{b} = \frac{1}{2}\left(\underline{B} + \overline{B}\right)$ *as the center of the interval* $\mathscr{B}$ *if* $\mathbf{0}_{n_x} \notin \mathscr{B}$, *and* $\tilde{b} = \mathbf{0}_{n_x}$ *if* $\mathbf{0}_{n_x} \in \mathscr{B}$. *Next define* $\tilde{\mathscr{B}} = \mathscr{B} - \tilde{b}$ *as the shifted interval vector which is guaranteed to contain* $\mathbf{0}_{n_x}$. *Then, with Taylor order* $r > \|\mathscr{A}\|_\infty (t_f - t_0) - 2$, *the interval*

$$H\left(\mathscr{X}_0, C_2(t_f - t_0) * \mathscr{X}_0 + C_3(t_f - t_0) * \tilde{b}\right)$$

$$+ C_3(t_f - t_0) * \tilde{\mathscr{B}} + C_4(t_f - t_0) * \mathscr{X}_0 + C_5(t_f - t_0) * \tilde{b}$$

is an over-approximation of the reachable tube $\bigcup_{t \in [t_0, t_f]} R\left(t; t_0, \mathscr{X}_0\right)$ *of the continuous-time affine system* (2.1).

Note that the function $H : \mathbb{I}^{n_x} \times \mathbb{I}^{n_x} \to \mathbb{I}^{n_x}$ denotes the interval hull of two intervals as defined in Sect. 2.1. The reachable tube defined in Proposition 2.3 is composed of three main elements. We first restrict the interval $\mathscr{B}$ to its center $\tilde{b}$ and take the interval hull of the interval of initial states $\mathscr{X}_0$ and the reachable set at the final time t_f computed as in Proposition 2.2. The second step is to expand this hull by $C_4(t_f - t_0) * \mathscr{X}_0 + C_5(t_f - t_0) * \tilde{b}$ to account for trajectories possibly going from the initial set to the final reachable set without staying within the interval hull. Finally, the term $C_3(t_f - t_0) * \tilde{\mathscr{B}}$ is added to account for the remaining values in $\mathscr{B}$ after shifting this interval by its center $\tilde{b}$. This last step can be done independently from the considerations on $\tilde{b}$ since we ensured that $\mathbf{0}_{n_x} \in \tilde{\mathscr{B}}$, which guarantees that $C_3(t_f - t_0) * \tilde{\mathscr{B}}$ is an expansion of the interval hull.

Discussion. The results presented in this section can be overly conservative. As in the case of discrete-time interval analysis in Sect. 2.2, this conservativeness is due to the multiple occurrences of the same intervals in the expressions of the reachable set and tube, and in particular, the power terms $\mathscr{A}^i$ in the Taylor expansion of the exponential function.

On the other hand, the results in this section are useful in another reachability method of this book in which an intermediate step requires the over-approximation of the reachable set and tube of a system's sensitivity equations whose affine dynamics match the conditions in this section. This method relying on the results of Propositions 2.2 and 2.3 is presented in Sects. 5.2 and 5.4 of Chap. 5.

Similar to the discrete-time case discussed in Sect. 2.2, there exist other methods relying on interval analysis for the reachability analysis of continuous-time systems. These methods are primarily based on Taylor series. While these alternatives are not presented in detail in this book, it should be noted that they allow for the consideration of a wider class of continuous-time systems (instead of only affine ones as in this section) with a reduced conservativeness of the over-approximations, but at the cost of greater computation times.

Further Reading

The definition of interval arithmetic operations in Sect. 2.1 and their use for interval reachability analysis of discrete-time system as in Sect. 2.2 can be found in textbooks on interval analysis, such as Jaulin et al. (2001) and Moore (1966).

For the reachability analysis of continuous-time affine systems, the results presented in Sect. 2.3 are from Althoff (2010) and Althoff et al. (2007). A variation of these results not presented in this book offering tighter bounds at the cost of additional computation are presented in Althoff et al. (2011).

Methods for the reachability analysis of more general continuous-time systems relying on Taylor expansions as mentioned at the end of Sect. 2.3 can be found, for example, in Chen et al. (2012), Raïssi et al. (2004) or the survey paper Nedialkov et al. (1999).

The use of redundant algebraic expressions of the same interval quantity to reduce the conservativeness of resulting interval has been used, for example, in Shen and Scott (2017).

References

Althoff M (2010) Reachability analysis and its application to the safety assessment of autonomous cars. Ph.D. thesis, Technische Universität München

Althoff M, Stursberg O, Buss M (2007) Reachability analysis of linear systems with uncertain parameters and inputs. In: 46th IEEE conference on decision and control. IEEE, pp 726–732

Althoff M, Le Guernic C, Krogh BH (2011) Reachable set computation for uncertain time-varying linear systems. In: Proceedings of the 14th international conference on hybrid systems: computation and control. ACM, pp 93–102

Chen X, Abraham E, Sankaranarayanan S (2012) Taylor model flowpipe construction for non-linear hybrid systems. In: 2012 IEEE 33rd real-time systems symposium. IEEE, pp 183–192

Jaulin L, Kieffer M, Didrit O, Walter E (2001) Applied interval analysis: with examples in parameter and state estimation, robust control and robotics, vol 1. Springer Science & Business Media

Moore RE (1966) Interval analysis, vol 4. Prentice-Hall Englewood Cliffs, NJ

Nedialkov NS, Jackson KR, Corliss GF (1999) Validated solutions of initial value problems for ordinary differential equations. Appl Math Comput 105(1):21–68

Raıssi T, Ramdani N, Candau Y (2004) Set membership state and parameter estimation for systems described by nonlinear differential equations. Automatica 40(10):1771–1777

Shen K, Scott JK (2017) Rapid and accurate reachability analysis for nonlinear dynamic systems by exploiting model redundancy. Comput Chem Eng 106:596–608

Chapter 3
Monotonicity

In this chapter, we present reachability methods taking advantage of the mathematical properties satisfied by monotone systems. We define the monotonicity property for discrete-time and continuous-time systems and present the associated reachability results in Sects. 3.1 and 3.2, respectively.

When applicable, these methods are guaranteed to provide a tight interval over-approximation of the reachable set, for both continuous-time and discrete-time systems. In addition, this approach is particularly efficient and has a low computational complexity since the over-approximations are fully defined from the evaluation of only two successors of the system. Finally, the monotonicity-based approach to reachability analysis forms the foundations for generalization of these efficient methods to any Lipschitz continuous system, presented in Chap. 4.

3.1 Discrete-Time Monotonicity

Monotonicity refers to a system whose trajectories preserve a partial order. A partial order on the Euclidean space $\mathbb{R}^{n_x}$ is defined as

$$x \preceq_x \hat{x} \iff \hat{x} - x \in K_x,$$

where $K_x \subseteq \mathbb{R}^{n_x}$ is a convex and pointed cone, thus satisfying $\alpha K_x \subseteq K_x$ for all $\alpha \geq 0$, $K_x + K_x \subseteq K_x$, and $K_x \cap (-K_x) = \{0\}$. For the input space $\mathbb{R}^{n_p}$, we similarly define the partial order $\preceq_p$ with respect to a cone K_p of $\mathbb{R}^{n_p}$ as

$$p \preceq_p \hat{p} \iff \hat{p} - p \in K_p.$$

© The Author(s), under exclusive license to Springer Nature Switzerland AG 2021
P.-J. Meyer et al., *Interval Reachability Analysis*,
SpringerBriefs in Control, Automation and Robotics,
https://doi.org/10.1007/978-3-030-65110-7_3

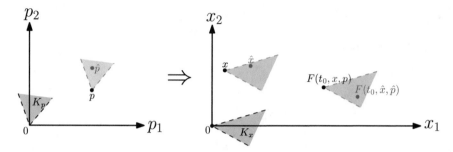

Fig. 3.1 Representation of the cones and induced partial orders in the input space (left-hand side) and state space (right-hand side). The discrete-time monotonicity in Definition 3.1 is illustrated through its preservation of the partial orders: if input $\hat{p}$ is in the translated cone K_p centered at input p, and initial state $\hat{x}$ is in the translated cone K_x centered at state x, then the successor state $F(t_0, \hat{x}, \hat{p})$ is in the translated cone K_x centered at successor state $F(t_0, x, p)$.

Definition 3.1 (*Discrete-time monotonicity*) A discrete-time system (1.1) is monotone with respect to partial orders $\preceq_x$ and $\preceq_p$ if the following implication holds for all $t_0 \in \mathbb{Z}$, $x, \hat{x} \in \mathbb{R}^{n_x}$ and $p, \hat{p} \in \mathbb{R}^{n_p}$

$$x \preceq_x \hat{x}, \ p \preceq_p \hat{p} \implies F(t_0, x, p) \preceq_x F(t_0, \hat{x}, \hat{p}).$$

This definition is illustrated in Fig. 3.1 in the case where $n_x = 2$ and $n_p = 2$. In the 2-dimensional input space, the considered cone is represented in the left-hand side of Fig. 3.1 and induces a partial order $p \preceq_p \hat{p}$ which can be interpreted as $\hat{p}$ belonging to the translated cone centered at p. In the 2-dimensional state space, the cone K_x is similarly represented in the right-hand side of Fig. 3.1. For a monotone system with respect to these partial orders, the vector field F preserves the order $\preceq_x$, meaning that $F(t_0, x, p) \preceq_x F(t_0, \hat{x}, \hat{p})$. This means that the successor $F(t_0, \hat{x}, \hat{p})$ belongs to the translated cone centered at $F(t_0, x, p)$, as illustrated in the right-hand side of Fig. 3.1.

While the general definition of monotonicity allows for arbitrary cones K_x and K_p to define the partial order as in Definition 3.1, in the remainder of this book, we focus on the subset of these cones defined by orthants of the Euclidean space. An orthant of $\mathbb{R}^{n_x}$ is the generalization of a half-space in $\mathbb{R}$ or a quadrant in $\mathbb{R}^2$: it is the Cartesian product of n_x half-spaces in $\mathbb{R}$, each equal to either $\mathbb{R}_+$ or $\mathbb{R}_-$. Then the partial order $\preceq_x$ induced by an orthant of $\mathbb{R}^{n_x}$ can be defined through the introduction of a Boolean vector $\varepsilon \in \{0, 1\}^{n_x}$ such that

$$x \preceq_x \hat{x} \iff (-1)^{\varepsilon}(\hat{x} - x) \geq 0,$$

using componentwise power function, multiplication, and inequality. In summary, $x \preceq_x \hat{x}$ implies that on each state dimension $i \in \{1, \ldots, n_x\}$, we have $x_i \leq \hat{x}_i$ if $\varepsilon_i = 0$ and $x_i \geq \hat{x}_i$ if $\varepsilon_i = 1$. Using such orthant-based partial orders is critical for working with intervals for both the initial set and the over-approximation of the reachable set in Problems 1.1 and 1.2. Relying on non-orthant cones would instead require the use of more complex set representations, corresponding to linear transformations (such as rotations and shear mappings) of axis-aligned hyper-rectangles, which would significantly increase the complexity of the numerical implementations to manipulate these sets.

Before providing the formal result to over-approximate the reachable set of a monotone system, we give a graphical interpretation of this over-approximation method.

Example 3.1 Consider a discrete-time system (1.2) with a 2-dimensional state space ($n_x = 2$) and a 1-dimensional input space ($n_p = 1$). Assume that the system is monotone with respect to the cones $K_x = \mathbb{R}_- \times \mathbb{R}_+$ and $K_p = \mathbb{R}_-$. The cone K_x induces a state partial order

$$x \preceq_x \hat{x} \quad \Longleftrightarrow \quad \begin{cases} x_1 \geq \hat{x}_1 \\ x_2 \leq \hat{x}_2 \end{cases},$$

which means that state $\hat{x}$ is *great than or equal to* x (with respect to $\preceq_x$) if $\hat{x}$ is above and to the left of x in $\mathbb{R}^2$. A state interval $[\underline{x}, \overline{x}] \subseteq \mathbb{R}^2$ can thus be reformulated to highlight the *lower and upper bounds* with respect to this partial order

$$[\underline{x}, \overline{x}] = \left\{ x \in \mathbb{R}^2 \;\middle|\; \begin{pmatrix} \overline{x}_1 \\ \underline{x}_2 \end{pmatrix} \preceq_x x \preceq_x \begin{pmatrix} \underline{x}_1 \\ \overline{x}_2 \end{pmatrix} \right\}.$$

Similarly, the cone $K_p = \mathbb{R}_-$ induces an input partial order corresponding to the reversed inequality $\preceq_p \equiv \geq$. The bounds of an input interval $[\underline{p}, \overline{p}] \subseteq \mathbb{R}$ with respect to this partial order are thus reversed as well

$$[\underline{p}, \overline{p}] = \left\{ p \in \mathbb{R} \mid \overline{p} \preceq_p p \preceq_p \underline{p} \right\}.$$

Therefore, for any initial state $x \in [\underline{x}, \overline{x}]$ and input $p \in [\underline{p}, \overline{p}]$, using Definition 3.1 twice, one obtains an interval over-approximation of the one-step reachable set

$$F\left(t_0, \begin{pmatrix} \overline{x}_1 \\ \underline{x}_2 \end{pmatrix}, \overline{p}\right) \preceq_x F(t_0, x, p) \preceq_x F\left(t_0, \begin{pmatrix} \underline{x}_1 \\ \overline{x}_2 \end{pmatrix}, \underline{p}\right),$$

which is also defined according to the partial order $\preceq_x$. The *lower bound* (respectively, *upper bound*) of the interval over-approximation is obtained by computing the successor of the system from the *lower bound* (respectively, *upper bound*) of the interval of initial states and with the *lower bound* (respectively, *upper bound*) of the interval of input values. These lower and upper bounds are all defined with respect to the partial orders $\preceq_x$ and $\preceq_p$, as illustrated in Fig. 3.2.

In addition, since the state and input vectors involved in the definition of the over-approximation belong to their respective intervals $[\underline{x}, \overline{x}]$ and $[\underline{p}, \overline{p}]$, then the

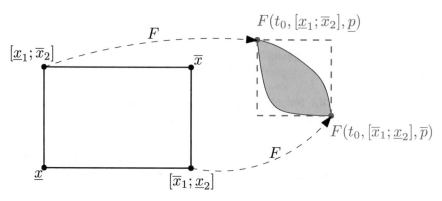

Fig. 3.2 Illustration of Example 3.1 to compute a tight interval over-approximation of the reachable set from the evaluation of only two successor states at the *lower and upper bounds* of the initial state and input intervals according to their respective partial orders

successors $F\left(t_0, \begin{pmatrix} \overline{x}_1 \\ \underline{x}_2 \end{pmatrix}, \overline{p}\right)$ and $F\left(t_0, \begin{pmatrix} \underline{x}_1 \\ \overline{x}_2 \end{pmatrix}, \underline{p}\right)$ are part of the reachable set $R\left(t_0, [\underline{x}, \overline{x}], [\underline{p}, \overline{p}]\right)$. It follows that the interval over-approximation is necessarily a tight over-approximation of the reachable set as in Definition 1.1. $\triangle$

We now formalize the main assumption and procedure illustrated in Example 3.1 to compute an interval over-approximation of the reachable set for a discrete-time monotone system.

Requirements and Limitations. For a discrete-time system (1.1), a characterization of the monotonicity property with respect to orthants of the state and input spaces is given based on the Jacobian matrices. This sufficient condition requires each entry of the Jacobian matrices $J_x(t, x, p)$ and $J_p(t, x, p)$ to be sign-stable over the considered ranges of state and input, with a specific sign structure formalized in the assumption below. We recall the notation of the entries of the Jacobian matrices as $J_{xij}(t_0, x, p) = \frac{\partial F_i(t_0, x, p)}{x_j}$ and $J_{p_{ik}}(t_0, x, p) = \frac{\partial F_i(t_0, x, p)}{p_k}$.

> **Assumption 3.1** There exist $\varepsilon = [\varepsilon_1; \ldots; \varepsilon_{n_x}] \in \{0, 1\}^{n_x}$ and $\delta = [\delta_1; \ldots; \delta_{n_p}] \in \{0, 1\}^{n_p}$ such that for all $x \in [\underline{x}, \overline{x}]$, $p \in [\underline{p}, \overline{p}]$, $i, j \in \{1, \ldots, n_x\}$ and $k \in \{1, \ldots, n_p\}$ we have
>
> $$(-1)^{\varepsilon_i + \varepsilon_j} J_{xij}(t_0, x, p) \geq 0, \qquad (-1)^{\varepsilon_i + \delta_k} J_{p_{ik}}(t_0, x, p) \geq 0.$$

The Boolean vectors ε and δ in Assumption 3.1 carry equivalent information to the orthants K_x and K_p or partial orders $\preceq_x$ and $\preceq_p$ used in Definition 3.1. In particular, if Assumption 3.1 is satisfied, then system (1.1) is monotone as in Definition 3.1 with respect to the partial orders

$$\begin{cases} x \preceq_x \hat{x} \iff \forall i \in \{1, \dots, n_x\}, \ (-1)^{\varepsilon_i}(\hat{x}_i - x_i) \geq 0, \\ p \preceq_p \hat{p} \iff \forall i \in \{1, \dots, n_p\}, \ (-1)^{\delta_i}(\hat{p}_i - p_i) \geq 0, \end{cases}$$

induced by orthants

$$K_x = K_{x1} \times \cdots \times K_{xn_x} \mid \forall i \in \{1, \dots, n_x\}, \ K_{xi} = \begin{cases} \mathbb{R}_+ \text{ if } \varepsilon_i = 0, \\ \mathbb{R}_- \text{ if } \varepsilon_i = 1, \end{cases}$$

$$K_p = K_{p1} \times \cdots \times K_{pn_p} \mid \forall i \in \{1, \dots, n_p\}, \ K_{pi} = \begin{cases} \mathbb{R}_+ \text{ if } \delta_i = 0, \\ \mathbb{R}_- \text{ if } \delta_i = 1. \end{cases}$$

Reachability Method. The main idea of reachability analysis based on the monotonicity property is to compute the successor states of (1.1) for only two state-input pairs $(x, p) \in [\underline{x}, \overline{x}] \times [\underline{p}, \overline{p}]$. The choice of these two pairs is done based on the Boolean vectors $\varepsilon = [\varepsilon_1; \dots; \varepsilon_{n_x}]$ and $\delta = [\delta_1; \dots; \delta_{n_p}]$ introduced in Assumption 3.1.

> **Proposition 3.1** *Under Assumption 3.1 and using componentwise multiplications with ε and δ, the interval*
>
> $$\Big[F\Big(t_0, \underline{x}\left(\mathbf{1}_{n_x} - \varepsilon\right) + \overline{x}\varepsilon, \ \underline{p}\left(\mathbf{1}_{n_p} - \delta\right) + \overline{p}\delta \Big),$$
> $$F\Big(t_0, \underline{x}\varepsilon + \overline{x}\left(\mathbf{1}_{n_x} - \varepsilon\right), \ \underline{p}\delta + \overline{p}\left(\mathbf{1}_{n_p} - \delta\right) \Big) \Big]$$
>
> *is an over-approximation of the one-step reachable set $R\Big(t_0, [\underline{x}, \overline{x}], [\underline{p}, \overline{p}] \Big)$ of (1.1) solving Problem 1.1.*

By construction, the two states $\underline{x}(\mathbf{1}_{n_x} - \varepsilon) + \overline{x}\varepsilon$ and $\underline{x}\varepsilon + \overline{x}(\mathbf{1}_{n_x} - \varepsilon)$ used to compute the over-approximation in Proposition 3.1 are diagonally opposite vertices of the interval $[\underline{x}, \overline{x}]$. The two inputs used in Proposition 3.1 are similarly diagonally opposite vertices of $[\underline{p}, \overline{p}]$. In the particular case where (1.1) is monotone with respect to the positive orthants $K_x = \mathbb{R}_+^{n_x}$ and $K_p = \mathbb{R}_+^{n_p}$, we have $\varepsilon = \mathbf{0}_{n_x}$ and $\delta = \mathbf{0}_{n_p}$ and we end up with the over-approximation interval $\Big[F\Big(t_0, \underline{x}, \underline{p}\Big), F\left(t_0, \overline{x}, \overline{p}\right) \Big]$ involving the classical lower and upper bounds of intervals $[\underline{x}, \overline{x}]$ and $[\underline{p}, \overline{p}]$ with respect to the partial orders defined by the componentwise inequalities.

Discussion. Relying on the monotonicity property is particularly powerful for reachability problems since the method described in Proposition 3.1 provides an over-approximation of the reachable set by only computing two successors of the discrete-time system (1.1). In addition, since both state-input pairs $(\underline{x}(\mathbf{1}_{n_x} - \varepsilon) + \overline{x}\varepsilon,$ $\underline{p}(\mathbf{1}_{n_p} - \delta) + \overline{p}\delta)$ and $(\underline{x}\varepsilon + \overline{x}(\mathbf{1}_{n_x} - \varepsilon), \underline{p}\delta + \overline{p}(\mathbf{1}_{n_p} - \delta))$ belong to $[\underline{x}, \overline{x}] \times [\underline{p}, \overline{p}]$, the interval in Proposition 3.1 is also guaranteed to be a tight interval over-approximation as in Definition 1.1.

3.2 Continuous-Time Monotonicity

For continuous-time systems, the definition of monotonicity and the associated reachability method is very similar to the discrete-time ones in Sect. 3.1, with only a few minor variations pointed out below.

Since the discrete-time reachability analysis in Sect. 3.1 focuses on the reachable set after a single time step, we only need to consider a single value for the input p over this time step. On the other hand, the continuous-time reachability analysis in this section is considered over a continuous time range $[t_0, t_f]$, on which inputs are allowed to be time-varying signals. We thus need to introduce the appropriate notion of partial order to compare two such inputs $\mathbf{p}, \hat{\mathbf{p}} : \mathbb{R} \to \mathbb{R}^{n_p}$,

$$\mathbf{p} \preceq_p \hat{\mathbf{p}} \iff \forall t \in \mathbb{R}, \ \hat{\mathbf{p}}(t) - \mathbf{p}(t) \in K_p,$$

where K_p is a cone of $\mathbb{R}^{n_p}$. Then the notion of monotonicity is defined as in the discrete-case of Definition 3.1 after replacing the discrete-time vector field F by the trajectory function Φ of the continuous-time system.

Definition 3.2 (*Continuous-time monotonicity*) A continuous-time system (1.2) is monotone with respect to partial orders $\preceq_x$ and $\preceq_p$ if the following implication holds for all $t_0, t_f \in \mathbb{R}$ with $t_f \geq t_0$, $x_0, \hat{x}_0 \in \mathbb{R}^{n_x}$ and $\mathbf{p}, \hat{\mathbf{p}} : [t_0, t_f] \to \mathbb{R}^{n_p}$

$$x_0 \preceq_x \hat{x}_0, \ \mathbf{p} \preceq_p \hat{\mathbf{p}} \implies \Phi(t_f; t_0, x_0, \mathbf{p}) \preceq_x \Phi(t_f; t_0, \hat{x}_0, \hat{\mathbf{p}}).$$

Since we only work with intervals as in Sect. 3.1, we formalize the main assumption and reachability result in the particular case where the cones inducing the partial orders in Definition 3.2 are orthants of the state and input spaces.

Requirements and Limitations. For a continuous-time system (1.2), the following condition characterizes the monotonicity with respect to orthants of the state and input spaces based on the sign structure of the Jacobian matrices.

Assumption 3.2 Given an invariant state space $X \subseteq \mathbb{R}^{n_x}$, there exist $\varepsilon = [\varepsilon_1; \ldots; \varepsilon_{n_x}] \in \{0, 1\}^{n_x}$ and $\delta = [\delta_1; \ldots; \delta_{n_p}] \in \{0, 1\}^{n_p}$ such that for all $t \in [t_0, t_f]$, $x \in X$, $p \in [\underline{p}, \overline{p}]$, $i, j \in \{1, \ldots, n_x\}$, $j \neq i$ and $k \in \{1, \ldots, n_p\}$ we have

$$(-1)^{\varepsilon_i + \varepsilon_j} J_{xij}(t, x, p) \geq 0, \qquad (-1)^{\varepsilon_i + \delta_k} J_{pik}(t, x, p) \geq 0.$$

Assumption 3.2 has two main differences from the Assumption 3.1 for discrete-time monotonicity. The first one is that the sign-stability and sign structure of the Jacobian matrices need to hold over the whole time range $[t_0, t_f]$ (instead of just t_0 for Assumption 3.1) and over an invariant X of the state space (instead of just the interval of initial states $\left[\underline{x}, \overline{x}\right]$ for Assumption 3.1) to ensure that these sign conditions are satisfied for all relevant trajectories of the continuous-time system. The second difference is that Assumption 3.2 does not have any conditions on the sign (or sign-stability) of the diagonal elements of the state Jacobian J_x.

The toolbox TIRA (Appendix B) automatically checks the satisfaction of Assumption 3.2 by translating the sign structure condition into a system of Boolean equations to be solved in the 2-element Galois Field GF(2).

Reachability Method. The over-approximation of the reachable set is then obtained in an identical way to the discrete-time case, simply by replacing both evaluations of the discrete-time successors F by evaluations of the continuous-time trajectories Φ at the final time t_f of the time range.

> **Proposition 3.2** *Under Assumption 3.2 and using componentwise multiplications with ε and δ, the interval*
>
> $$\left[\Phi\left(t_f; t_0, \underline{x}\left(\mathbf{1}_{n_x} - \varepsilon\right) + \overline{x}\varepsilon, \underline{p}\left(\mathbf{1}_{n_p} - \delta\right) + \overline{p}\delta\right),\right.$$
> $$\left.\Phi\left(t_f; t_0, \underline{x}\varepsilon + \overline{x}\left(\mathbf{1}_{n_x} - \varepsilon\right), \underline{p}\delta + \overline{p}\left(\mathbf{1}_{n_p} - \delta\right)\right)\right]$$
>
> *is an over-approximation of the reachable set $R\left(t_f; t_0, \left[\underline{x}, \overline{x}\right], \left[\underline{p}, \overline{p}\right]\right)$ of (1.2) solving Problem 1.2.*

Discussion. The reachability result in Proposition 3.2 based on continuous-time monotonicity has the same advantages as its discrete-time counterpart: the interval over-approximation is guaranteed to be tight and its computation only requires the evaluation of two finite-time successors of the continuous-time system (1.2).

On the other hand, both the discrete-time method in Sect. 3.1 and the continuous-time one in this section are only applicable to monotone systems, which thus need to satisfy the sign-stability and sign structure conditions on the Jacobian matrices as in Assumption 3.1 and 3.2, respectively. These conditions are strong limitations since the majority of systems do not have sign-stable Jacobian matrices or the required sign structure. Therefore, in Chap. 4, we present generalizations of these monotonicity-based methods to general nonlinear systems, using a notion called mixed monotonicity. The cost of these generalizations is that the obtained over-approximations may not always be tight.

Further Reading

The definition of discrete-time monotone systems and their infinitesimal characterization as in Assumption 3.1 are presented in Hirsch and Smith (2005). For continuous-time monotonicity, these definitions and characterizations are given in Angeli and Sontag (2003). This monotonicity property is used for interval reachability analysis in Moor and Raisch (2002) and Ramdani et al. (2010).

Although there is a rich literature on monotone dynamical systems, we do not detail it here as we discuss only definitions and properties relevant to reachability analysis. To learn more about the general topic of monotonicity, the reader is referred to the book Hirsch and Smith (2006), survey papers Smith (1988), Hirsch and Smith (2005) and their own references.

Another related work relies on similar notions of bounding functions and Metzler matrices (a particular case of Assumption 3.2 with all off-diagonal elements being nonnegative) to create interval observers for nonlinear systems Efimov et al. (2013).

References

Angeli D, Sontag ED (2003) Monotone control systems. IEEE Trans Autom Control 48(10):1684–1698

Efimov D, Raïssi T, Chebotarev S, Zolghadri A (2013) Interval state observer for nonlinear time varying systems. Automatica 49(1):200–205

Hirsch MW, Smith H (2005) Monotone maps: a review. J Differ Equ Appl 11(4–5):379–398

Hirsch MW, Smith H (2006) Monotone dynamical systems. In: Handbook of differential equations: ordinary differential equations, vol 2. Elsevier, pp 239–357

Moor T, Raisch J (2002) Abstraction based supervisory controller synthesis for high order monotone continuous systems. In: Modelling, analysis, and design of hybrid systems. Springer, pp 247–265

Ramdani N, Meslem N, Candau Y (2010) Computing reachable sets for uncertain nonlinear monotone systems. Nonlinear Anal Hybrid Syst 4(2):263–278

Smith HL (1988) Systems of ordinary differential equations which generate an order preserving flow. A survey of results. SIAM Rev 30(1):87–113

Chapter 4
Mixed Monotonicity

Despite their efficiency and guaranteed tightness, the applicability of the over-approximation methods in Chap. 3 is limited to the class of monotone systems. In this chapter, we provide generalizations of monotonicity-based methods by considering a wider class of *mixed-monotone* systems. In particular, we show that the mixed-monotonicity property is satisfied by any Lipschitz continuous nonlinear system, thus making the associated reachability methods broadly applicable.

The first method in Sect. 4.1 is based on the notion of mixed monotonicity for continuous-time systems and is a generalization of the approach relying on continuous-time monotonicity in Sect. 3.2. The discrete-time version of mixed monotonicity and its associated reachability method is presented in Sect. 4.2, which generalizes Sect. 3.1. A third approach defined as *sampled-data mixed monotonicity*, which applies the discrete-time method in Sect. 4.2 to the sampled-data version of a continuous-time system, is presented in Chap. 5.

4.1 Continuous-Time Mixed Monotonicity

A continuous-time system (1.2) is called mixed monotone if it can be embedded in an auxiliary system which is monotone.

Definition 4.1 The continuous-time system (1.2) with vector field $f : \mathbb{R} \times \mathbb{R}^{n_x} \times \mathbb{R}^{n_p} \to \mathbb{R}^{n_x}$ is *mixed monotone* if there exists a *decomposition function* $g : \mathbb{R} \times \mathbb{R}^{n_x} \times \mathbb{R}^{n_p} \times \mathbb{R}^{n_x} \times \mathbb{R}^{n_p} \to \mathbb{R}^{n_x}$ such that for all $t \in \mathbb{R}$, $x, \hat{x} \in \mathbb{R}^{n_x}$ and $p, \hat{p} \in \mathbb{R}^{n_p}$, the following three conditions hold

- g is increasing in its first state-input pair:
  for all $i, j \in \{1, \ldots, n_x\}$, $j \neq i$ and $k \in \{1, \ldots, n_p\}$, we have

$$\frac{\partial g_i(t, x, p, \hat{x}, \hat{p})}{\partial x_j} \geq 0, \qquad \frac{\partial g_i(t, x, p, \hat{x}, \hat{p})}{\partial p_k} \geq 0,$$

- g is decreasing in its second state-input pair:
  for all $i, j \in \{1, \ldots, n_x\}$ and $k \in \{1, \ldots, n_p\}$, we have

$$\frac{\partial g_i(t, x, p, \hat{x}, \hat{p})}{\partial \hat{x}_j} \leq 0, \qquad \frac{\partial g_i(t, x, p, \hat{x}, \hat{p})}{\partial \hat{p}_k} \leq 0,$$

- f is embedded in the diagonal of g:

$$g(t, x, p, x, p) = f(t, x, p).$$

This decomposition function implies that the new dynamical system evolving in $\mathbb{R}^{2n_x}$

$$\begin{pmatrix} \dot{x} \\ \dot{\hat{x}} \end{pmatrix} = h(t, x, p, \hat{x}, \hat{p}) = \begin{pmatrix} g(t, x, p, \hat{x}, \hat{p}) \\ g(t, \hat{x}, \hat{p}, x, p) \end{pmatrix}, \tag{4.1}$$

is monotone with respect to the orthant $\mathbb{R}^{n_x}_+ \times \mathbb{R}^{n_x}_-$ in its state space $\mathbb{R}^{2n_x}$ and the orthant $\mathbb{R}^{n_p}_+ \times \mathbb{R}^{n_p}_-$ in its input space $\mathbb{R}^{2n_p}$ (i.e. it satisfies Assumption 3.2 with $\varepsilon = [\mathbf{0}_{n_x}; \mathbf{1}_{n_x}]$ and $\delta = [\mathbf{0}_{n_p}; \mathbf{1}_{n_p}]$). The reachability analysis of the original system (1.2) can then be deduced from the reachability analysis method of Chap. 3 applied to the monotone system (4.1) and combined with the embedding of f in g as in Definition 4.1.

Requirements and Limitations. For a continuous-time system (1.2), one condition for the applicability of this method is to find two matrices $L_x \in \mathbb{R}^{n_x \times n_x}$ and $L_p \in \mathbb{R}^{n_x \times n_p}$ such that $J_x(t, x, p) + L_x$ and $J_p(t, x, p) + L_p$ are both sign-stable matrices over the considered ranges of time, state, and input. This sign-stability condition is to be interpreted in a componentwise fashion, meaning that each element of $J_x(t, x, p) + L_x$ (apart from its diagonal) and of $J_p(t, x, p) + L_p$ should keep a constant sign while the time, state, and input vary.

Assumption 4.1 Given an invariant state space $X \subseteq \mathbb{R}^{n_x}$, there exists $L_x \in \mathbb{R}^{n_x \times n_x}$ such that for all $i, j \in \{1, \ldots, n_x\}$ with $j \neq i$ we have either

$$J_{xij}(t, x, p) + L_{xij} \geq 0, \qquad \forall t \in [t_0, t_f], x \in X, p \in \left[\underline{p}, \overline{p}\right]$$

or

$$J_{xij}(t, x, p) + L_{xij} \leq 0, \qquad \forall t \in [t_0, t_f], x \in X, p \in \left[\underline{p}, \overline{p}\right],$$

and there exists $L_p \in \mathbb{R}^{n_x \times n_p}$ such that for all $i \in \{1, \ldots, n_x\}$ and $k \in \{1, \ldots, n_p\}$ we have either

$$J_{p_{ik}}(t, x, p) + L_{p_{ik}} \geq 0, \qquad \forall t \in [t_0, t_f], x \in X, p \in \left[\underline{p}, \overline{p}\right]$$

or

$$J_{p_{ik}}(t, x, p) + L_{p_{ik}} \leq 0, \qquad \forall t \in [t_0, t_f], x \in X, p \in \left[\underline{p}, \overline{p}\right].$$

This is a much less restrictive assumption than Assumption 3.2, since we do not require any sign structure for the sign-stable matrices, and the sign-stability conditions on J_x and J_p from Assumption 3.2 are relaxed by allowing for constant matrices L_x and L_p to obtain the sign-stability on $J_x + L_x$ and $J_p + L_p$. Note that if both Jacobian matrices are already sign-stable, then picking $L_x = \mathbf{0}_{n_x \times n_x}$ and $L_p = \mathbf{0}_{n_x \times n_p}$ results in the satisfaction of Assumption 4.1.

The conditions in Assumption 4.1 can also be interpreted in terms of boundedness of the Jacobian matrices. Indeed, the addition of the constant matrices L_x and L_p aims to shift the range of possible values of the Jacobian matrices such that for each element of the matrices, this shifted range is fully contained either in the positive half-space or in the negative half-space. As a result, Assumption 4.1 can be satisfied if the range of values taken by each element of J_x and J_p is either upper bounded or lower bounded. A more detailed description of how to create matrices L_x and L_p based on the initial knowledge of bounds on the Jacobian matrices is provided in Appendix A.

Reachability Method. Assumption 4.1 gives sufficient conditions under which a decomposition function g satisfying Definition 4.1 can be constructed for the continuous-time system (1.2). Each component $g_i : \mathbb{R} \times \mathbb{R}^{n_x} \times \mathbb{R}^{n_p} \times \mathbb{R}^{n_x} \times \mathbb{R}^{n_p} \to \mathbb{R}$ with $i \in \{1, \ldots, n_x\}$ of the decomposition function is defined as

$$g_i(t, x, p, \hat{x}, \hat{p}) = f_i(t, \xi^i, \pi^i) + |L_{xi*}|(x - \hat{x}) + |L_{p_{i*}}|(p - \hat{p}), \qquad (4.2)$$

where state $\xi^i \in \mathbb{R}^{n_x}$ and input $\pi^i \in \mathbb{R}^{n_p}$ are defined below based on the sign of the elements of the shifting matrices L_x and L_p in Assumption 4.1, and $L_{xi*} \in \mathbb{R}^{1 \times n_x}$ and $L_{p_{i*}} \in \mathbb{R}^{1 \times n_p}$ are the row vectors representing the ith row of matrices L_x and L_p, respectively. The elements of $\xi^i = \left[\xi_1^i; \ldots; \xi_{n_x}^i\right]$ match the elements of either x or $\hat{x}$ depending on the signs in the shifting matrix L_x, where for all $j \in \{1, \ldots, n_x\}$ we define

$$\xi_j^i = \begin{cases} x_j & \text{if } L_{xij} \geq 0, \\ \hat{x}_j & \text{if } L_{xij} < 0. \end{cases} \qquad (4.3)$$

Note that if the shifting matrix L_x in Assumption 4.1 is obtained according to the recommendations in Appendix A, then (4.3) leads to $\xi_i^i = x_i$ since we have $L_{xii} = 0$ by default. The input vector $\pi^i = \left[\pi_1^i; \ldots; \pi_{n_p}^i\right]$ is similarly obtained based on the signs of the elements in matrix L_p, where for all $k \in \{1, \ldots, n_p\}$ we define

$$\pi^i_k = \begin{cases} p_k & \text{if } L_{p_{ik}} \geq 0, \\ \hat{p}_k & \text{if } L_{p_{ik}} < 0. \end{cases} \tag{4.4}$$

We can then confirm that the decomposition function in (4.2) indeed satisfies the three conditions from Definition 4.1. First, it correctly embeds the vector field $(g(t, x, p, x, p) = f(t, x, p))$ since when $(\hat{x}, \hat{p}) = (x, p)$, (4.3) and (4.4) give $\xi^i = x$ and $\pi^i = p$ for all $i \in \{1, \ldots, n_x\}$. In addition, it can be shown that due to the choice of L_x and L_p in Assumption 4.1, all non-diagonal elements of the partial derivative $\frac{\partial g}{\partial x}(t, x, p, \hat{x}, \hat{p})$ and all elements of $\frac{\partial g}{\partial p}(t, x, p, \hat{x}, \hat{p})$ are nonnegative, and that all elements of $\frac{\partial g}{\partial \hat{x}}(t, x, p, \hat{x}, \hat{p})$ and $\frac{\partial g}{\partial \hat{p}}(t, x, p, \hat{x}, \hat{p})$ are nonpositive. This means that for all $i \in \{1, \ldots, n_x\}$, the function $g_i(t, x, p, \hat{x}, \hat{p})$ is indeed increasing with all x_j (with $j \neq i$) and all p_k, and decreasing with all $\hat{x}_j$ and all $\hat{p}_k$. This also confirms that the auxiliary system (4.1) is monotone as in Assumption 3.2, due to the resulting sign structure of its Jacobian matrices.

The result below is obtained by first applying Proposition 3.2 to over-approximate the reachable set of (4.1), and then using the embedding of f in g to deduce a reachability result solving Problem 1.2 for the original non-monotone system (1.2). Given an initial state $[x_0; \hat{x}_0] \in \mathbb{R}^{2n_x}$ at time $t_0 \in \mathbb{R}$ and a constant input $[p; \hat{p}] \in \mathbb{R}^{2n_p}$, the successor of (4.1) at time $t_f \geq t_0$ is denoted as $\Phi^h(t_f; t_0, x_0, p, \hat{x}_0, \hat{p}) \in \mathbb{R}^{2n_x}$. The first and last n_x components of Φ^h are denoted as $\Phi^h_{1\ldots n_x}$ and $\Phi^h_{n_x+1\ldots 2n_x}$, respectively. The presented approach thus results in an over-approximation of the reachable set of (1.2) from the evaluation of a single successor of the $2n_x$-dimensional system (4.1).

> **Proposition 4.1** *Under Assumption 4.1 and definitions* (4.1)–(4.4), *the interval*
>
> $$\left[\Phi^h_{1\ldots n_x}\left(t_f; t_0, \underline{x}, \underline{p}, \overline{x}, \overline{p}\right), \Phi^h_{n_x+1\ldots 2n_x}\left(t_f; t_0, \underline{x}, \underline{p}, \overline{x}, \overline{p}\right) \right]$$
>
> *is an over-approximation of the reachable set* $R\left(t_f; t_0, [\underline{x}, \overline{x}], [\underline{p}, \overline{p}]\right)$ *of* (1.2) *solving Problem 1.2.*

In Proposition 4.1, we are able to obtain reachability result for (1.2) using only the successor Φ^h of the duplicated system (4.1) because of the embedding of f in g. In addition, the symmetry of the dynamics of (4.1) implies that $\Phi^h_{1\ldots n_x}(t_f; t_0, \overline{x}, \overline{p}, \underline{x}, \underline{p}) = \Phi^h_{n_x+1\ldots 2n_x}(t_f; t_0, \underline{x}, p, \overline{x}, \overline{p})$, which is the reason why a single evaluation of Φ^h is sufficient for the result on (1.2) in Proposition 4.1, even though the full reachability analysis on (4.1) would require two evaluations.

Discussion. As mentioned in Appendix A, an equivalent formulation of Assumption 4.1 is that all non-diagonal elements of the state Jacobian matrix and all elements of the input Jacobian matrix are either lower bounded or upper bounded over the considered ranges of time, state, and inputs. On the other hand, it can be shown that any Lipschitz continuous nonlinear system necessarily has bounded Jacobian matrices over bounded ranges of time, state, and inputs. Therefore, the reachability result presented in Proposition 4.1 is applicable to any Lipschitz continuous nonlinear system.

Although any Lipschitz continuous system (1.2) admits shifting matrices L_x and L_p satisfying Assumption 4.1, the computation of an interval over-approximation as in Proposition 4.1 requires the explicit knowledge of these shifting matrices. Thus, the application of the constructive result presented in this section relies on the ability of the user to compute bounds on the Jacobian matrices (which are then used as in Appendix A to deduce the shifting matrices L_x and L_p).

Since the continuous-time monotonicity property described by Assumption 3.2 is a particular case of Assumption 4.1, the general reachability result in Proposition 4.1 can also be applied to monotone systems. In addition, it is proved in Meyer et al. (2019) that for a continuous-time monotone system as in Assumption 3.2, the interval over-approximations obtained from Propositions 3.2 and 4.1 are the same and correspond to the unique tight interval over-approximation of the reachable set. Indeed, a monotone system necessarily has sign-stable Jacobian matrices (apart from the diagonal of the state Jacobian matrix), which means that defining the shifting matrices as $L_x = \mathbf{0}_{n_x \times n_x}$ and $L_p = \mathbf{0}_{n_x \times n_p}$ is sufficient for the satisfaction of Assumption 4.1. Then, from equations (4.2)–(4.4), it follows that the decomposition function is equal to the monotone vector field ($g(t, x, p, \hat{x}, \hat{p}) = f(t, x, p)$) and that the larger system (4.1) merely contains two independent copies of (1.2).

In terms of computation time, the method from Proposition 3.2 for monotone systems and the more general result in Proposition 4.1 have comparable complexity for the computation of an interval over-approximation. Indeed, the first method based on monotonicity relies on the computation of two finite-time successors of the n_x-dimensional system (1.2), while the second and more general method using mixed monotonicity only requires the computation of a single finite-time successor but for a larger $2n_x$-dimensional system (4.1).

4.2 Discrete-Time Mixed Monotonicity

The generalization of discrete-time monotonicity from Sect. 3.1 into a notion of discrete-time mixed monotonicity is achieved similarly to their continuous-time counterparts in Sect. 4.1. The resulting approach provides a very general reachability method applicable to any Lipschitz continuous discrete-time system.

Requirements and Limitations. The reachability method presented in this section is applicable to any discrete-time system as in (1.1) for which we can find two matrices $L_x \in \mathbb{R}^{n_x \times n_x}$ and $L_p \in \mathbb{R}^{n_x \times n_p}$ such that $J_x(t_0, x, p) + L_x$ and $J_p(t_0, x, p) + L_p$ are both sign-stable matrices over the considered ranges of state and input.

Assumption 4.2 There exists $L_x \in \mathbb{R}^{n_x \times n_x}$ such that for all $i, j \in \{1, \ldots, n_x\}$ we have either

$$J_{xij}(t_0, x, p) + L_{xij} \geq 0, \qquad \forall x \in \left[\underline{x}, \overline{x}\right], p \in \left[\underline{p}, \overline{p}\right]$$

or

$$J_{xij}(t_0, x, p) + L_{xij} \leq 0, \qquad \forall x \in \left[\underline{x}, \overline{x}\right], p \in \left[\underline{p}, \overline{p}\right],$$

and there exists $L_p \in \mathbb{R}^{n_x \times n_p}$ such that for all $i \in \{1, \ldots, n_x\}$ and $k \in \{1, \ldots, n_p\}$ we have either

$$J_{pik}(t_0, x, p) + L_{pik} \geq 0, \qquad \forall x \in \left[\underline{x}, \overline{x}\right], p \in \left[\underline{p}, \overline{p}\right]$$

or

$$J_{pik}(t_0, x, p) + L_{pik} \leq 0, \qquad \forall x \in \left[\underline{x}, \overline{x}\right], p \in \left[\underline{p}, \overline{p}\right].$$

Assumption 4.2 has two differences from Assumption 4.1 for continuous-time mixed monotonicity. The first difference is that Assumption 4.2 only needs to hold at the initial time t_0 (while Assumption 4.1 covers the whole time range $[t_0, t_f]$) and on the interval of initial states $\left[\underline{x}, \overline{x}\right]$ (unlike Assumption 4.1 where an invariant set in the state space $X \subseteq \mathbb{R}^{n_x}$ needs to be found). The second difference is that Assumption 4.2 includes a condition on the diagonal elements of the state Jacobian, while Assumption 4.1 does not. This means that for discrete-time systems, it may be necessary to set the diagonal elements of the shifting matrix L_x to non-zero values.

Similar to the continuous-time case, Assumption 4.2 is equivalent to having all elements of both Jacobian matrices to be either lower bounded or upper bounded. The values of the constant shifting matrices L_x and L_p in Assumption 4.2 can then be deduced from these Jacobian bounds by following the instructions in Appendix A for all elements of the Jacobian matrices (including the diagonal of J_x). Finally, the Lipschitz continuity of the discrete-time vector field F in (1.1) is a sufficient condition for the satisfaction of Assumption 4.2. Therefore, the reachability method presented below is broadly applicable.

Reachability Method. The formulation of the reachability method solving Problem 1.1 for discrete-time systems satisfying Assumption 4.2 is very similar to its continuous-time counterpart presented in Sect. 4.1. The decomposition function $g : \mathbb{Z} \times \mathbb{R}^{n_x} \times \mathbb{R}^{n_p} \times \mathbb{R}^{n_x} \times \mathbb{R}^{n_p} \to \mathbb{R}^{n_x}$ is defined as in (4.2)–(4.4) after replacing the continuous-time vector field f by the discrete-time one F, that is for all $i \in \{1, \ldots, n_x\}$ the ith component of the decomposition function is

$$g_i(t_0, x, p, \hat{x}, \hat{p}) = F_i(t_0, \xi^i, \pi^i) + |L_{xi*}|(x - \hat{x}) + |L_{pi*}|(p - \hat{p}), \qquad (4.5)$$

where $L_{xi*} \in \mathbb{R}^{1 \times n_x}$ and $L_{pi*} \in \mathbb{R}^{1 \times n_p}$ are the row vectors representing the ith row of matrices L_x and L_p, respectively, and state $\xi^i = \left[\xi_1^i; \ldots; \xi_{n_x}^i\right] \in \mathbb{R}^{n_x}$ and input $\pi^i = \left[\pi_1^i; \ldots; \pi_{n_p}^i\right] \in \mathbb{R}^{n_p}$ are defined such that for all $j \in \{1, \ldots, n_x\}$ and $k \in \{1, \ldots, n_p\}$,

$$\xi_j^i = \begin{cases} x_j & \text{if } L_{xij} \geq 0, \\ \hat{x}_j & \text{if } L_{xij} < 0, \end{cases} \qquad \pi_k^i = \begin{cases} p_k & \text{if } L_{pik} \geq 0, \\ \hat{p}_k & \text{if } L_{pik} < 0. \end{cases} \qquad (4.6)$$

Since we now consider discrete-time systems, the definition of a discrete-time version of the duplicated system (4.1) is not necessary. Instead, for the statement of the discrete-time reachability result, it is sufficient to replace the successors of this duplicated system Φ^h in Proposition 4.1 by evaluations of the decomposition function g.

Proposition 4.2 *Under Assumption 4.2 and definitions (4.5)–(4.6), the interval*

$$\left[g\left(t_0, \underline{x}, \underline{p}, \overline{x}, \overline{p}\right), g\left(t_0, \overline{x}, \overline{p}, \underline{x}, \underline{p}\right) \right]$$

is an over-approximation of the one-step reachable set $R\left(t_0, \left[\underline{x}, \overline{x}\right], \left[\underline{p}, \overline{p}\right]\right)$ of (1.1) solving Problem 1.1.

Discussion. The general mixed-monotonicity result in Proposition 4.2 encompasses as particular cases the reachability analysis based on discrete-time monotone systems in Proposition 3.1, as well as the reachability result relying on sign-stable Jacobian matrices as in Coogan and Arcak (2015) (corresponding to the particular case where $L_x = \mathbf{0}_{n_x \times n_x}$ and $L_p = \mathbf{0}_{n_x \times n_p}$ in Assumption 4.2). In addition, in both these particular cases (discrete-time monotonicity, or sign-stability of both Jacobian matrices) the interval computed in Proposition 4.2 is guaranteed to be the unique tight interval over-approximation of the reachable set. In comparison with the results for continuous-time systems, tightness is also guaranteed for continuous-time monotone systems as in Assumption 3.2, but not in the case where the Jacobian matrices of the continuous-time system are sign-stable. This tightness difference for the sign-stable cases comes from the fact that the continuous-time reachability analysis is done over a continuous time range during which the conservativeness may accumulate when the system is not monotone, while the discrete-time reachability analysis is done over a single discrete step which does not allow this accumulation of conservativeness.

In terms of computation time, this method has a comparable complexity to both discrete-time and continuous-time monotonicity results in Propositions 3.1 and 3.2, and to continuous-time mixed monotonicity in Proposition 4.1. Indeed, the over-approximation in Proposition 4.2 requires two evaluations of the n_x-dimensional decomposition function g, while Propositions 3.1 and 3.2 rely on two evaluations of the successor for the n_x-dimensional system, and Proposition 4.1 only uses a single successor of the $2n_x$-dimensional system (4.1).

Further Reading

The idea of decomposing the increasing and decreasing components of a non-monotone system and embedding them into a larger monotone system was first introduced in Gouzé and Hadeler (1994), and is applicable to both continuous-time (Enciso et al. 2006) and discrete-time systems (Smith 2006). It was then brought to the field of reachability analysis in Coogan and Arcak (2015).

A first infinitesimal characterization of continuous-time systems satisfying such mixed-monotonicity property was given in Coogan and Arcak (2016), Coogan et al. (2016), where a constructive result for the decomposition of non-monotone dynamics is provided for any system whose Jacobian matrices are sign-stable. This sufficient condition is thus significantly more general than the one for monotonicity since it removes the sign structure condition from Assumption 3.2. The results from Coogan et al. (2016) were then generalized in Yang et al. (2019) to any nonlinear system whose Jacobian matrices are bounded, later used in Meyer and Dimarogonas (2019) for a reachability analysis problem. Finally, a relaxation of the boundedness conditions from Yang et al. (2019) to reduce the conservativeness of the reachability analysis results was provided in Meyer et al. (2019).

For discrete-time systems, the first characterization of mixed monotonicity based on the sign-stability of the Jacobian matrices was introduced in Coogan and Arcak (2015). This characterization was then generalized in Meyer et al. (2018) to any discrete-time system with bounded Jacobian matrices as described in Sect. 4.2. The paper Coogan and Arcak (2015) also provides examples of discrete-time systems whose mixed monotonicity can be deduced without studying their Jacobian matrices. Finally, a recent paper has proved that for any mixed-monotone function, there exists a decomposition function (implicitly defined as the solution of an optimization problem) resulting in a tight over-approximation (Yang and Ozay 2019).

Other relevant works include (Angeli and Sontag 2013) defining a graphical representation of the decomposition into increasing and decreasing behaviors of a non-monotone system with a sign-stable Jacobian matrix, and Angeli et al. (2014) considering closed-loop systems with controllers defined as a mix of both positive feedback and negative feedback.

References

Angeli D, Sontag ED (2013) Behavior of responses of monotone and sign-definite systems. In: Mathematical systems theory, pp 51–64

Angeli D, Enciso GA, Sontag ED (2014) A small-gain result for orthant-monotone systems under mixed feedback. Syst Control Lett 68:9–19

Coogan S, Arcak M (2015) Efficient finite abstraction of mixed monotone systems. In: 18th international conference on hybrid systems: computation and control, pp 58–67

Coogan S, Arcak M (2016) Stability of traffic flow networks with a polytree topology. Automatica 66:246–253

Coogan S, Arcak M, Kurzhanskiy AA (2016) Mixed monotonicity of partial first-in-first-out traffic flow models. In: 55th ieee conference on decision and control, pp 7611–7616

Enciso GA, Smith HL, Sontag ED (2006) Nonmonotone systems decomposable into monotone systems with negative feedback. J Differ Equ 224(1):205–227

Gouzé JL, Hadeler KP (1994) Monotone flows and order intervals. Nonlinear World 1:23–34

Meyer PJ, Dimarogonas DV (2019) Hierarchical decomposition of LTL synthesis problem for nonlinear control systems. IEEE Trans Autom Control 64(11):4676–4683

Meyer PJ, Coogan S, Arcak M (2018) Sampled-data reachability analysis using sensitivity and mixed-monotonicity. IEEE Control Syst Lett 2(4):761–766

Meyer PJ, Devonport A, Arcak M (2019) TIRA: toolbox for interval reachability analysis. In: Proceedings of the 22nd ACM international conference on hybrid systems: computation and control. ACM, pp 224–229

Smith HL (2006) The discrete dynamics of monotonically decomposable maps. J Math Biol 53(4):747

Yang L, Ozay N (2019) Tight decomposition functions for mixed monotonicity. In: IEEE 58th conference on decision and control. IEEE, pp 5318–5322

Yang L, Mickelin O, Ozay N (2019) On sufficient conditions for mixed monotonicity. In: IEEE transactions on automatic control

Chapter 5
Sampled-Data Mixed Monotonicity

This chapter applies the discrete-time mixed-monotonicity method from Sect. 4.2 to
the sampled-data version of a continuous-time system, thus overcoming the conservativeness of the continuous-time approach in Sect. 4.1. The main difficulty arising
from this approach is that the boundedness assumption on the Jacobian matrices of
the discrete-time system in Sect. 4.2 is converted into a boundedness assumption
on the *sensitivity matrices* of the continuous-time system. Since these sensitivity
matrices are partial derivatives of the continuous-time trajectories whose closed-
form expression is usually unknown, the satisfaction of this assumption requires
significant effort.

The reachability analysis based on the notion of sampled-data mixed monotonicity is first presented in Sect. 5.1 under the assumption that bounds on the sensitivity
matrices are known. The following three sections then provide methods to compute
such bounds: Sect. 5.2 using the interval analysis from Chap. 2 on the dynamics of
the sensitivity matrix; Sect. 5.3 approximating this bound through a sampling-based
approach; and Sect. 5.4 presenting a hybrid version of the previous two methods
resulting in a sound over-approximation with a tunable tradeoff between its conservativeness and computational complexity.

5.1 Sampled-Data Mixed Monotonicity from Bounded Sensitivity

As discussed in Chap. 4, the reachability method based on discrete-time mixed monotonicity in Proposition 4.2 has a significant advantage compared to the one relying on
continuous-time mixed monotonicity in Proposition 4.1. Indeed, while both methods
result in a tight interval over-approximation when the system satisfies the associated
monotonicity condition, only the discrete-time version preserves this tightness in the
more general case where the Jacobian matrices are sign-stable.

P.-J. Meyer et al., *Interval Reachability Analysis*,
SpringerBriefs in Control, Automation and Robotics,
https://doi.org/10.1007/978-3-030-65110-7_5

In addition, the conservativeness of the over-approximation using the continuous-time mixed-monotonicity method increases with the size of the time range $[t_0, t_f]$ defined in the reachability Problem 1.2. On the other hand, such an increase in conservativeness with time does not appear in the discrete-time methods since Problem 1.1 only considers the reachable set after a single step of the discrete-time system.

The advantages of the discrete-time method motivate an alternative approach to continuous-time systems where we make use of a sampled-data model with a sampling period of $t_f - t_0$, and then apply the discrete-time mixed-monotonicity results from Proposition 4.2.

Requirements and Limitations. This method is applicable to continuous-time systems as in (1.2), with the additional restriction that the input function needs to be constant over the considered time range $[t_0, t_f]$ of the reachability problem. This restriction arises because we want to define the sampled-data version of this continuous-time system as a discrete-time system $x^+ = F(t_0, x_0, p) = \Phi(t_f; t_0, x_0, \mathbf{p})$. This equality is properly defined only when $\mathbf{p}(t) = p$ for all $t \in [t_0, t_f]$ since a discrete-time system fully abstracts the notion of time between the initial state x_0 and its one-step successor $x^+ = F(t_0, x_0, p)$, making it unable to account for time-varying input signals between these two instants. As a consequence, this constant input can be modeled as a dynamical system $\dot{p} = 0$ with an initialization in the interval $\left[\underline{p}, \overline{p}\right] \in \mathbb{R}^{n_p}$ to describe the uncertainty of its constant value. Therefore, to reduce notation, we consider that this input is already included in the state of the continuous-time system (1.2), which can thus be simplified into the system $\dot{x} = f(t, x)$ where $f_i(t, x) = 0$ for each state component x_i corresponding to a constant input.

In this method, we define the sampled-data version of this continuous-time system as a discrete-time system

$$x^+ = F(t_0, x_0) = \Phi(t_f; t_0, x_0), \tag{5.1}$$

where the discrete-time successor corresponds to the successor of the continuous-time system at the final time t_f. Then the objective is to apply the discrete-time mixed-monotonicity reachability method from Sect. 4.2 to this system. Since this method relies on the Jacobian matrix $J_x(t_0, x_0) = \frac{\partial F(t_0, x_0)}{\partial x_0}$ of the discrete-time system, the translation of the conditions from Assumption 4.2 to the sampled-data continuous-time system involves the partial derivative of the sampled-data successor function

$$S_x(t_f; t_0, x_0) = \frac{\partial \Phi(t_f; t_0, x_0)}{\partial x_0}. \tag{5.2}$$

This matrix is referred to as the sensitivity matrix of the continuous-time system, since it represents the sensitivity of the continuous-time successors at the final time t_f with respect to variations of the initial state.

Assumption 5.1 There exists $L_x \in \mathbb{R}^{n_x \times n_x}$ such that for all $i, j \in \{1, \ldots, n_x\}$ we have either

$$S_{xij}(t_f; t_0, x_0) + L_{xij} \geq 0, \qquad \forall x_0 \in \left[\underline{x}, \overline{x}\right]$$

or

$$S_{xij}(t_f; t_0, x_0) + L_{xij} \leq 0, \qquad \forall x_0 \in \left[\underline{x}, \overline{x}\right].$$

Note that Assumption 5.1 does not need to hold for the sensitivity values at all times $t \in [t_0, t_f]$, but only for its final value at time t_f. Similar to both mixed-monotonicity methods in Chap. 4, if we know sensitivity bounds $\left[\underline{S_x}, \overline{S_x}\right]$ such that $S_x(t_f; t_0, x_0) \in \left[\underline{S_x}, \overline{S_x}\right]$ for all $x_0 \in \left[\underline{x}, \overline{x}\right]$, then we can use Appendix A to deduce the shifting matrix L_x in Assumption 5.1.

While bounds on the Jacobian matrices of either continuous-time or discrete-time systems can be computed for most systems by studying their vector field, bounding the sensitivity matrix of a continuous-time system is significantly more challenging. Indeed, the sensitivity matrix (5.2) depends on the solution map Φ, whose closed-form expression cannot be obtained explicitly for any nontrivial continuous-time system. For now, we assume that sensitivity bounds (or the equivalent shifting matrix L_x as in Assumption 5.1) are known. Then, in Sects. 5.2–5.4, we provide three methods to compute such bounds.

Reachability Method. The formulation of the reachability method solving Problem 1.2 for the continuous-time system satisfying Assumption 5.1 then consists of applying the discrete-time mixed-monotonicity result from Proposition 4.2 to the sampled-data system (5.1). The first step is to create a decomposition function $g : \mathbb{R} \times \mathbb{R}^{n_x} \times \mathbb{R}^{n_x} \to \mathbb{R}^{n_x}$ whose ith component with $i \in \{1, \ldots, n_x\}$ is

$$g_i(t_0, x, \hat{x}) = \Phi_i(t_f; t_0, \xi^i) + |L_{xi*}|(x - \hat{x}), \tag{5.3}$$

where $L_{xi*} \in \mathbb{R}^{1 \times n_x}$ is the ith row of matrix L_x and state $\xi^i = \left[\xi_1^i; \ldots; \xi_{n_x}^i\right] \in \mathbb{R}^{n_x}$ is defined such that for all $j \in \{1, \ldots, n_x\}$,

$$\xi_j^i = \begin{cases} x_j & \text{if } L_{xij} \geq 0, \\ \hat{x}_j & \text{if } L_{xij} < 0. \end{cases} \tag{5.4}$$

Then, the reachability Problem 1.2 for the continuous-time system is solved by computing only two evaluations of the decomposition function.

Proposition 5.1 *Under Assumption 5.1 and definitions (5.3)–(5.4), the interval*

$$\left[g\left(t_0, \underline{x}, \overline{x}\right), g\left(t_0, \overline{x}, \underline{x}\right)\right]$$

is an over-approximation of the reachable set $R\left(t_f; t_0, \left[\underline{x}, \overline{x}\right]\right)$ of (1.2) solving Problem 1.2.

Discussion. Similarly to the discrete-time mixed-monotonicity method in Sect. 4.2, if the sensitivity matrix $S_x(t_f; t_0, x_0)$ is sign-stable over the set of initial states $\left[\underline{x}, \overline{x}\right]$ (meaning that the shifting matrix is $L_x = \mathbf{0}_{n_x \times n_x}$), then Proposition 5.1 results in a tight interval over-approximation of the reachable set.

In terms of computation time, Proposition 5.1 has the same complexity as the discrete-time method in Proposition 4.2, since both only require two evaluations of the decomposition function. On the other hand, this complexity only takes into account the reachability analysis when the shifting matrix L_x in Assumption 5.1 is known or, equivalently, bounds on the sensitivity matrix are available. The computation of bounds for this sensitivity matrix is discussed in Sects. 5.2–5.4, and depending on the chosen approach, may induce a significantly higher computational complexity.

5.2 Sensitivity Bounds from Interval Analysis

As is well known in sensitivity analysis, taking the time derivative of (5.2) and applying the chain rule results in a continuous-time system describing the time variations of the sensitivity matrix

$$\dot{S}_x(t; t_0, x_0) = J_x(t, \Phi(t; t_0, x_0)) * S_x(t; t_0, x_0), \tag{5.5}$$

where $J_x(t, x) = \frac{\partial f(t,x)}{\partial x}$ is the Jacobian matrix of the continuous-time system (1.2). Since at the initial time t_0 we have $\Phi(t_0; t_0, x_0) = x_0$, then the definition of the sensitivity matrix implies that the initial condition of this system is the identity matrix

$$S_x(t_0; t_0, x_0) = I_{n_x}.$$

We then note that this sensitivity system (5.5) is linear in its state $S_x(t; t_0, x_0)$. Therefore, the objective of the approach presented in this section is to apply the interval analysis method in Sect. 2.3 to this system, whose reachable set over-approximation computed as in Proposition 2.2 will provide interval bounds of the sensitivity matrix at the final time t_f.

Requirements and Limitations. In Sect. 2.3, the reachability analysis of the affine system (2.1) requires Assumption 2.1, which stipulates that $A(t)$ and $B(t)$ be bounded by intervals. To transpose this assumption to the particular case of the sensitivity system, we first note that (5.5) does not contain the additive term $B(t)$ which can thus be bounded by a singleton with the zero matrix $\mathbf{0}_{n_x \times n_x}$. Then the only remaining requirement is to provide interval bounds for matrix $A(t)$ corresponding to the state Jacobian in (5.5).

Assumption 5.2 Given an invariant state space $X \subseteq \mathbb{R}^{n_x}$, there exists $\left[\underline{J_x}, \overline{J_x}\right] \in \mathbb{I}^{n_x \times n_x}$ such that for all $t \in [t_0, t_f]$ and $x \in X$ we have

$$J_x(t, x) \in \left[\underline{J_x}, \overline{J_x}\right].$$

Method. Similarly to Sect. 2.3, we consider the shorthand notation $\mathcal{J}_x = \left[\underline{J_x}, \overline{J_x}\right]$ and we denote the infinity norm of an interval matrix as $\|\mathcal{J}_x\|_\infty = \left\|\max\left(\left|\underline{J_x}\right|, \left|\overline{J_x}\right|\right)\right\|_\infty$ using componentwise functions for the absolute value and maximum. Next, for a Taylor order $r \in \mathbb{N}$ satisfying $r > \|\mathcal{J}_x\|_\infty (t_f - t_0) - 2$, we introduce the function $C_1 : \mathbb{R}_+ \to \mathbb{I}^{n_x \times n_x}$

$$C_1(\tau) = \left[-\mathbf{1}_{n_x \times n_x}, \mathbf{1}_{n_x \times n_x}\right] * \frac{(\|\mathcal{J}_x\|_\infty \tau)^{r+1}}{(r+1)!} \frac{r+2}{r+2 - \|\mathcal{J}_x\|_\infty \tau}.$$

Then with $X_0 = \{I_{n_x}\}$ and $\mathcal{B} = \{0_{n_x \times n_x}\}$, Proposition 2.2 provides the following interval over-approximation for the sensitivity matrix.

Proposition 5.2 *Under Assumption 5.2 and with Taylor order $r > \|\mathcal{J}_x\|_\infty (t_f - t_0) - 2$, the interval*

$$\sum_{i=0}^{r} \frac{(\mathcal{J}_x(t_f - t_0))^i}{i!} + C_1(t_f - t_0)$$

is an over-approximation of the set $\left\{S_x(t_f; t_0, x_0) \mid x_0 \in \left[\underline{x}, \overline{x}\right]\right\}$ of possible sensitivity values at the final time t_f.

Discussion. Compared to the next method in Sect. 5.3, the main strength of the result in Proposition 5.2 is that the computed interval is guaranteed to over-approximate the set of possible sensitivity values. This means that this interval satisfies Assumption 5.1 and that in turn, the main reachability method in Sect. 5.2 and Proposition 5.1 provides a guaranteed over-approximation of the reachable set of the continuous-time system (1.2).

Another advantage is that this method has a fairly low computational complexity in most cases, since the computation of the interval over-approximation is only based on $3r + 2$ interval arithmetic operations. On the other hand, such low computational complexity is not always guaranteed. When the time range $[t_0, t_f]$ is wide and the bounds of the Jacobian matrix $\left[\underline{J_x}, \overline{J_x}\right]$ have large absolute values, then the minimal Taylor order r for Proposition 5.2 to hold becomes very large and the corresponding $3r + 2$ interval arithmetic operations can require a significant computation time.

Finally, the main shortcoming of this approach is that it tends to provide overly conservative over-approximations of the set of sensitivity values, which then propagates into conservativeness of the reachability results in Proposition 5.1.

5.3 Sensitivity Bounds from Sampling

In this section, we consider a simulation-based approach which approximates the sensitivity bounds by evaluating the sensitivity matrix $S_x(t_f; t_0, x_0)$ for a set of sampled initial states. This can then be followed by an optional falsification step to iteratively improve the accuracy of the obtained bounds by solving an optimization problem.

As discussed further at the end of this section, this approach results in much more accurate approximations of the set of sensitivity values than the interval analysis approach in Sect. 5.2, but the obtained sensitivity interval is not guaranteed to be an over-approximation of the set of sensitivity values.

Requirements and Limitations. One of the main advantages of this approach is that its application does not rely on additional assumptions. The only requirement to initialize this method is to define a non-empty finite subset of initial states

$$X \subseteq \left[\underline{x}, \overline{x}\right]$$

to be used for the evaluation of the sensitivity matrix. The number and positions of these sampled initial states can be chosen randomly or through the definition of a uniform grid of the interval $\left[\underline{x}, \overline{x}\right]$. Chapter 7 presents an alternative Monte Carlo strategy that samples the interval according to some probability distribution and gives probabilistic guarantees on the resulting approximation.

The accuracy of the approximation of the sensitivity bounds increases with the number of samples for which the sensitivity matrix is evaluated. On the other hand, the computation time necessary for these numerical evaluations is also directly proportional to the number of samples. The cardinality of the chosen sampling set can thus be used to trade computation time for accuracy of the results.

Method. The first step is to compute a numerical evaluation of the sensitivity matrix (5.2) for each sampled initial state $x_0 \in X$ in the finite sampling set. Note that Assumption 5.1 requires the evaluation of the sensitivity only at the final time t_f, which can be obtained through any classical numerical method (such as Euler or Runge–Kutta) depending on the allowed numerical error and computation time.

An alternative for the evaluation of these sensitivity matrices is to solve numerically the linear dynamical system of the sensitivity (5.5) for each $x_0 \in X$ and evaluate the solution at the final time t_f. However, this alternative tends to require longer computation times without guaranteeing lower numerical errors.

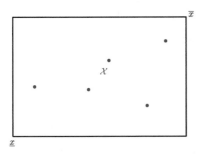 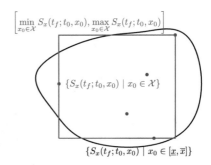

Fig. 5.1 Sampling-based interval approximation of the set of sensitivity values. *Left (in the state space $\mathbb{R}^{n_x}$)*: interval of initial states $[\underline{x}, \overline{x}]$ and finite set of samples X. *Right (in the sensitivity space $\mathbb{R}^{n_x \times n_x}$)*: evaluation of the sensitivity value for the sampled states, interval hull of these evaluations as in (5.6), and true set of sensitivity values for comparison

From a finite number of evaluations of the sensitivity matrix, we can thus approximate the set of possible sensitivity values $\{S_x(t_f; t_0, x_0) \mid x_0 \in [\underline{x}, \overline{x}]\}$ by the interval defined as the componentwise extrema of the sampled evaluations

$$\left[\min_{x_0 \in X} S_x(t_f; t_0, x_0), \max_{x_0 \in X} S_x(t_f; t_0, x_0) \right]. \tag{5.6}$$

The procedure leading to the interval approximation of the sensitivity set in (5.6) is sketched in Fig. 5.1. On the left-hand side, we start from a finite set of samples X in the interval of initial states $[\underline{x}, \overline{x}]$. Then moving to the sensitivity space $\mathbb{R}^{n_x \times n_x}$ (sketched in 2D in the right-hand side of Fig. 5.1 for visualization), we evaluate numerically the sensitivity $S_x(t_f; t_0, x_0)$ for each sampled state $x_0 \in X$ and take the interval hull of these evaluations corresponding to the componentwise min and max in (5.6). The set of all sensitivity values $\{S_x(t_f; t_0, x_0) \mid x_0 \in [\underline{x}, \overline{x}]\}$ is represented for comparison. In particular, this illustration shows that the interval in (5.6) may be neither an over-approximation nor an under-approximation of this set.

Since the approximation (5.6) is the interval hull of a strict subset $\{S_x(t_f; t_0, x_0) \mid x_0 \in X\}$ of the true set of sensitivity values $\{S_x(t_f; t_0, x_0) \mid x_0 \in [\underline{x}, \overline{x}]\}$, it is impossible to guarantee that this interval over-approximates the set of all possible sensitivity values. On the other hand, increasing the number of sampled initial states in $X \subseteq [\underline{x}, \overline{x}]$ is likely to reduce the volume of the set difference

$$\{S_x(t_f; t_0, x_0) \mid x_0 \in [\underline{x}, \overline{x}]\} \setminus \left[\min_{x_0 \in X} S_x(t_f; t_0, x_0), \max_{x_0 \in X} S_x(t_f; t_0, x_0) \right].$$

In particular, if the sampling is done through any probability distribution whose support is the whole interval $[\underline{x}, \overline{x}]$, then interval (5.6) converges to the unique tight

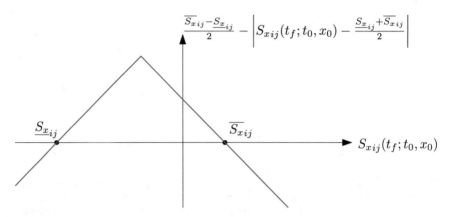

Fig. 5.2 Cost function to be minimized over the set of initial states for element (i, j) of the sensitivity matrix $S_x(t_f; t_0, x_0)$

interval over-approximation of the sensitivity set $\{S_x(t_f; t_0, x_0) \mid x_0 \in [\underline{x}, \overline{x}]\}$ when the number of samples goes to infinity.

Another way to improve the quality of the interval approximation (5.3) without adding more samples is through a *falsification* approach. The objective is to look for initial states in $[\underline{x}, \overline{x}]$ whose sensitivity evaluation does not belong to the interval approximation (5.6) obtained above through sampling. Denoting as $[\underline{S_x}, \overline{S_x}]$ the interval in (5.6), the falsification is attempted by solving the optimization problem

$$\min_{x_0 \in [\underline{x}, \overline{x}]} \left(\min_{i,j} \left(\frac{\overline{S}_{xij} - \underline{S}_{xij}}{2} - \left| S_{xij}(t_f; t_0, x_0) - \frac{\underline{S}_{xij} + \overline{S}_{xij}}{2} \right| \right) \right).$$

In this problem, for each element (i, j) of the sensitivity matrix, we minimize over the set of initial states the cost function in Fig. 5.2, which is positive if $S_{xij}(t_f; t_0, x_0) \in \left[\underline{S}_{x_{ij}}, \overline{S}_{x_{ij}} \right]$ and negative otherwise. Since we also take the minimum of this cost function over all pairs $(i, j) \in \{1, \ldots, n_x\}^2$, we conclude that if the optimization problem returns a negative result for some minimizer x_0^*, then we have $S_x(t_f; t_0, x_0^*) \notin \left[\underline{S}_x, \overline{S}_x \right]$. In such case, we update the sensitivity bounds in (5.6)

$$\underline{S}_x \leftarrow \min \left(\underline{S}_x, S_x(t_f; t_0, x_0^*) \right),$$
$$\overline{S}_x \leftarrow \max \left(\overline{S}_x, S_x(t_f; t_0, x_0^*) \right),$$

using componentwise min and max operators. We can then repeat this falsification process for a new iteration using the updated bounds if we want to further improve the approximation. On the other hand, since this optimization problem is only able to find local minima, finding a positive minimum does not mean that the updated sensitivity bounds over-approximate the set of all possible sensitivity values.

Discussion. This simulation-based approach to approximating the sensitivity bounds has two main advantages. The first one is that its application does not require any assumption on the system other than having constant inputs as mentioned in the main reachability result of Sect. 5.1. In particular, this means that unlike the interval analysis approach in Sect. 5.2, the user does not need to compute bounds on the Jacobian matrices for the use of the sampling and falsification methods.

The second strong advantage compared to the interval analysis approach in Sect. 5.2 is that it produces an interval approximation of the set of sensitivity values with significantly reduced conservativeness. More specifically, for any number of sampled initial states and falsification iterations used, the computed interval approximation of the sensitivity bounds is necessarily contained in the interval hull (or tight interval over-approximation) of the set of all possible sensitivity values $\{S_x(t_f; t_0, x_0) \mid x_0 \in [\underline{x}, \overline{x}]\}$. The distance between the computed approximation and the true tight interval over-approximation can then be reduced either by including the optional falsification step or by considering additional sampled states.

On the other hand, the main drawback of this approach is that the computed interval approximation can never be guaranteed to be an over-approximation of the set of sensitivity values. This is true even in the case where the interval approximation is widened through the falsification method. For the reachability analysis of safety-critical systems, it is thus recommended to instead consider the sound approaches in Sects. 5.2 and 5.4 whose resulting sensitivity bounds are guaranteed to be over-approximations, although sometimes more conservative. For other systems, we can also look for probabilistic guarantees on the interval approximation by applying the Monte Carlo method described in Chap. 7.

The second drawback of this simulation-based approach is its computational complexity. Indeed, the number of samples has a strong influence on the quality of the obtained approximation. If, for example, we consider a classical grid sampling of the initial state interval $[\underline{x}, \overline{x}] \subseteq \mathbb{R}^{n_x}$, then the number of samples is exponential in the state dimension n_x. The optional falsification steps may induce significant additional computation time, since solving the optimization problem requires evaluating the sensitivity matrix several times.

5.4   Sensitivity Bounds: Hybrid Approach

The method based on interval analysis in Sect. 5.2 provides a sound over-approximation, usually obtained with low computation times, but tends to be very conservative. The simulation-based approach in Sect. 5.3 results in better approximations of the set of sensitivity values, but without soundness guarantees and at the cost of large computational complexity. The third approach proposed in this section aims to combine elements of both previous methods to obtain a sound over-approximation of the set of sensitivity values with the ability to tune the tradeoff between complexity and conservativeness.

Before providing an outline of this method, we need to introduce new notations specific to this section. We first recall that the derivative of a scalar differentiable function $f : \mathbb{R}^n \to \mathbb{R}$ with respect to its vector argument is defined as the row vector

$$\frac{\partial f(x)}{\partial x} = \left(\frac{\partial f(x)}{\partial x_1} \quad \cdots \quad \frac{\partial f(x)}{\partial x_n} \right).$$

Then for a functional matrix $A : \mathbb{R}^n \to \mathbb{R}^{p \times q}$, the derivative $\frac{\partial A(x)}{\partial x} \in \mathbb{R}^{p \times nq}$ is defined as a $p \times q$ block matrix where each element $A_{ij}(x) \in \mathbb{R}$ of $A(x) \in \mathbb{R}^{p \times q}$ is replaced by the row vector of its derivative $\frac{\partial A_{ij}(x)}{\partial x} \in \mathbb{R}^{1 \times n}$:

$$
\begin{aligned}
\frac{\partial A(x)}{\partial x} &= \begin{pmatrix} \frac{\partial A_{11}(x)}{\partial x} & \cdots & \frac{\partial A_{1q}(x)}{\partial x} \\ \vdots & \ddots & \vdots \\ \frac{\partial A_{p1}(x)}{\partial x} & \cdots & \frac{\partial A_{pq}(x)}{\partial x} \end{pmatrix} \\
&= \begin{pmatrix} \frac{\partial A_{11}(x)}{\partial x_1} & \cdots & \frac{\partial A_{11}(x)}{\partial x_n} & \cdots & \frac{\partial A_{1q}(x)}{\partial x_1} & \cdots & \frac{\partial A_{1q}(x)}{\partial x_n} \\ \vdots & & \vdots & \vdots & \vdots & & \vdots \\ \frac{\partial A_{p1}(x)}{\partial x_1} & \cdots & \frac{\partial A_{p1}(x)}{\partial x_n} & \cdots & \frac{\partial A_{pq}(x)}{\partial x_1} & \cdots & \frac{\partial A_{pq}(x)}{\partial x_n} \end{pmatrix}.
\end{aligned}
$$

This notation ensures that we only work with 2-dimensional matrices, instead of matrices with more than two dimensions for which cumbersome matrix product definitions would need to be introduced.

The interval analysis approach in Sect. 5.2 assumes the boundedness of the first-order Jacobian matrix $J_x(t, x) = \frac{\partial f(t,x)}{\partial x}$ to deduce an over-approximation of the first-order sensitivity matrix (5.2). In this section, we also rely on the second-order Jacobian matrix

$$J_{xx}(t, x) = \frac{\partial J_x(t, x)}{\partial x} \in \mathbb{R}^{n_x \times n_x^2}$$

and the second-order sensitivity matrix

$$S_{xx}(t; t_0, x_0) = \frac{\partial S_x(t; t_0, x_0)}{\partial x_0} \in \mathbb{R}^{n_x \times n_x^2}.$$

In Sect. 5.2, we used the definition of the first-order sensitivity matrix and the chain rule to obtain a continuous-time linear system (5.5) describing the dynamics of S_x. Similarly, we can use variations of the product rule and chain rule adapted to functional matrices to obtain a continuous-time affine system describing the dynamics of the second-order sensitivity matrix

$$
\begin{aligned}
\dot{S}_{xx}(t; t_0, x_0) =\, & J_x(t, \Phi(t; t_0, x_0)) * S_{xx}(t; t_0, x_0) \qquad\qquad\qquad (5.7) \\
& + J_{xx}(t, \Phi(t; t_0, x_0)) * (S_x(t; t_0, x_0) \otimes S_x(t; t_0, x_0)),
\end{aligned}
$$

with initial condition $S_{xx}(t_0; t_0, x_0) = \mathbf{0}_{n_x \times n_x^2}$.

The approach presented in this section proceeds in three steps. We first over-approximate the reachable tube (over the whole time range $[t_0, t_f]$) of the first-order sensitivity matrix using interval analysis as in Proposition 2.3. We then use this result to over-approximate the reachable set (at the final time t_f) of the second-order sensitivity matrix using Proposition 2.2. Finally, this result is combined with a finite number of evaluations of the first-order sensitivity matrix over a grid of sampled initial states in $[\underline{x}, \overline{x}]$, which provides a guaranteed over-approximation of the reachable set (at the final time t_f) of the first-order sensitivity matrix. The granularity of the sampling grid can then be used to tune the computational complexity and conservativeness of this final result.

Requirements and Limitations. Both the first-order and second-order Jacobian matrices appear in the dynamics of the first-order sensitivity matrix (5.5) and second-order sensitivity matrix (5.7). Therefore, to apply the interval analysis results from Sect. 2.3, we need interval bounds for both Jacobian matrices.

Assumption 5.3 Given an invariant state space $X \subseteq \mathbb{R}^{n_x}$, there exist $\left[\underline{J_x}, \overline{J_x} \right] \in \mathbb{I}^{n_x \times n_x}$ and $\left[\underline{J_{xx}}, \overline{J_{xx}} \right] \in \mathbb{I}^{n_x \times n_x^2}$ such that for all $t \in [t_0, t_f]$ and $x \in X$ we have

$$J_x(t, x) \in \left[\underline{J_x}, \overline{J_x} \right],$$
$$J_{xx}(t, x) \in \left[\underline{J_{xx}}, \overline{J_{xx}} \right].$$

Although $S_x(t; t_0, x_0)$ also appears in the dynamics of the second-order sensitivity matrix (5.7), the user does not need to provide bounds on the first-order sensitivity matrix since such bounds are computed through the first step of the method below.

Method. As mentioned above, the first two steps of this method consist of applying the interval analysis results from Sect. 2.3 to both sensitivity systems: first applying Proposition 2.3 to (5.5) for the first-order sensitivity, and then applying Proposition 2.2 to (5.7) for the second-order sensitivity. Since both the sensitivity systems (5.5) and (5.7) have their state multiplied by the same matrix $J_x(t, \Phi(t; t_0, x_0))$, we recall some notation from Sect. 2.3 to be used in both steps. We first take the shorthand notation $\mathcal{J}_x = \left[\underline{J_x}, \overline{J_x} \right]$ and we denote its infinity norm as $\|\mathcal{J}_x\|_\infty = \left\| \max\left(\left| \underline{J_x} \right|, \left| \overline{J_x} \right| \right) \right\|_\infty$ using componentwise functions for the absolute value and maximum. Next, for a Taylor order $r \in \mathbb{N}$ satisfying $r > \|\mathcal{J}_x\|_\infty (t_f - t_0) - 2$, we introduce the functions $C_1, C_2, C_3, C_4 : \mathbb{R}_+ \to \mathbb{I}^{n_x \times n_x}$

$$C_1(\tau) = \left[-\mathbf{1}_{n_x \times n_x}, \mathbf{1}_{n_x \times n_x} \right] * \frac{(\|\mathcal{J}_x\|_\infty \tau)^{r+1}}{(r+1)!} \frac{r+2}{r+2 - \|\mathcal{J}_x\|_\infty \tau},$$

$$C_2(\tau) = \sum_{i=0}^{r} \frac{(\mathcal{J}_x \tau)^i}{i!} + C_1(\tau),$$

$$C_3(\tau) = \sum_{i=0}^{r} \frac{\mathcal{J}_x^i \tau^{i+1}}{(i+1)!} + C_1(\tau)\tau,$$

$$C_4(\tau) = \left[\sum_{i=2}^{r} \left(i^{\frac{-i}{i-1}} - i^{\frac{-1}{i-1}}\right) \frac{(\mathcal{J}_x \tau)^i}{i!}, \mathbf{0}_{n_x \times n_x}\right] + C_1(\tau).$$

In the first step, we over-approximate the reachable tube over the whole time range $[t_0, t_f]$ for the first-order sensitivity matrix (5.5) using interval analysis as in Proposition 2.3. Using the notation of Sect. 2.3, the linear system (5.5) describing the first-order sensitivity is characterized by the interval bounds of its parameters $\mathcal{A} = \mathcal{J}_x = \left[\underline{J_x}, \overline{J_x}\right]$ and $\mathcal{B} = \left\{\mathbf{0}_{n_x \times n_x}\right\}$ and the interval of initial states $X_0 = \left\{I_{n_x}\right\}$.

Lemma 5.1 *Under Assumption 5.3 and denoting as H the interval hull of two interval matrices, the interval*

$$\left[\underline{S_x^{RT}}, \overline{S_x^{RT}}\right] = H\left(\{I_n\}, C_2(t_f - t_0)\right) + C_4(t_f - t_0) \tag{5.8}$$

is an over-approximation of the reachable tube $\left\{S_x(t; t_0, x_0) \mid x_0 \in \left[\underline{x}, \overline{x}\right], t \in [t_0, t_f]\right\}$ of first-order sensitivity values over the time range $[t_0, t_f]$.

The second step of this approach is to over-approximate the reachable set at the final time t_f for the second-order sensitivity (5.7) using interval analysis as in Proposition 2.2. Combining (5.8) with the bounds on the first-order and second-order Jacobian matrices from Assumption 5.3, we can indeed bound both parameters of the affine system of the second-order sensitivity (5.7). Using the notation of Sect. 2.3, the interval bounds of these system parameters are defined as $\mathcal{A} = \mathcal{J}_x = \left[\underline{J_x}, \overline{J_x}\right]$ and $\mathcal{B} = \left[\underline{J_{xx}}, \overline{J_{xx}}\right] * \left(\left[\underline{S_x^{RT}}, \overline{S_x^{RT}}\right] \otimes \left[\underline{S_x^{RT}}, \overline{S_x^{RT}}\right]\right)$, and the interval of initial states is the singleton $X_0 = \left\{\mathbf{0}_{n_x \times n_x^2}\right\}$.

Lemma 5.2 *Under Assumption 5.3 leading to the bounds on the reachable tube of the first-order sensitivity (5.8), the interval*

$$\left[\underline{S_{xx}}, \overline{S_{xx}}\right] = C_3(t_f - t_0) * \left[\underline{J_{xx}}, \overline{J_{xx}}\right] * \left(\left[\underline{S_x^{RT}}, \overline{S_x^{RT}}\right] \otimes \left[\underline{S_x^{RT}}, \overline{S_x^{RT}}\right]\right) \tag{5.9}$$

is an over-approximation of the reachable set $\left\{S_{xx}(t_f; t_0, x_0) \mid x_0 \in \left[\underline{x}, \overline{x}\right]\right\}$ of second-order sensitivity values at the final time t_f.

The final step consists of combining the second-order sensitivity bounds (5.9) with evaluations of $S_x(t_f; t_0, x_0)$ at sampled initial states, in order to obtain an interval over-approximation of the reachable set at the final time t_f for the first-order sensitivity matrix. We first introduce the notion of dispersion of a finite subset of initial states, where the infinity norm of a state $x \in \mathbb{R}^{n_x}$ is defined as $\|x\|_\infty = \max_{i \in \{1,\dots,n_x\}} |x_i|$.

Definition 5.1 Given a finite set $X \subseteq [\underline{x}, \overline{x}]$, the dispersion of X in $[\underline{x}, \overline{x}]$ is defined as:

$$d(X) = \sup_{x \in [\underline{x}, \overline{x}]} \min_{y \in X} \|x - y\|_\infty \in \mathbb{R}.$$

The dispersion $d(X)$ thus takes smaller values when the sample states in X are well scattered in the interval $[\underline{x}, \overline{x}]$.

The results below are valid for any choice of the sampling set $X \subseteq [\underline{x}, \overline{x}]$, as long as its dispersion can be either computed exactly or upper-bounded. This task is challenging for most choices of sampling set, whose dispersion can usually only be lower-bounded when the state dimension n_x is more than one. We thus focus on the particular case of the sampling set defined as a uniform grid on $[\underline{x}, \overline{x}]$ as in Fig. 5.3 (left), for which the dispersion can be computed exactly.

Lemma 5.3 Let X be defined as a uniform grid in $[\underline{x}, \overline{x}]$ with $a \in \mathbb{N}$ elements per dimension (i.e., containing a^{n_x} sample states) and such that on each dimension $i \in \{1, \dots, n_x\}$ the samples are separated by $\frac{\overline{x}_i - \underline{x}_i}{a}$ and the first sample is shifted of $\frac{\overline{x}_i - \underline{x}_i}{2a}$ from $\underline{x}_i$. Then the dispersion of X is given by

$$d(X) = \frac{\|\overline{x} - \underline{x}\|_\infty}{2a}.$$

Finally, we combine the evaluations of the first-order sensitivity matrix $S_x(t_f; t_0, x_0)$ at each sampled initial state $x_0 \in X$ with the second-order sensitivity bounds (5.9) to deduce an interval over-approximation of the set of first-order sensitivity values at the final time t_f.

> **Proposition 5.3** *Under Assumption 5.3 leading to the bounds on the second-order sensitivity (5.9) and given a finite set $X \subseteq [\underline{x}, \overline{x}]$ of sampled initial states, define $M \in \mathbb{R}^{n_x \times n_x}$ as*
>
> $$M = \max \left(|\underline{S_{xx}}|, |\overline{S_{xx}}| \right) * \left(I_{n_x} \otimes \left(\mathbf{1}_{n_x} * d(X) \right) \right),$$
>
> *using componentwise absolute value and max operators. Then the interval*
>
> $$\left[\min_{x_0 \in X} \left(S_x(t_f; t_0, x_0) \right) - M, \max_{x_0 \in X} \left(S_x(t_f; t_0, x_0) \right) + M \right],$$

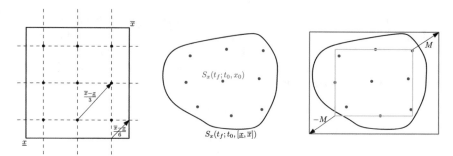

Fig. 5.3 The procedure in Proposition 5.3. *Left*: sampling of the interval of initial states $\left[\underline{x}, \overline{x}\right]$ into a uniform grid of $a = 3$ samples per dimension. *Middle*: evaluation of the first-order sensitivity $S_x(t_f; t_0, x_0)$ for each sampled initial state x_0, and representation of the whole reachable set $\left\{S_x(t_f; t_0, x_0) \mid x_0 \in \left[\underline{x}, \overline{x}\right]\right\}$. *Right*: interval hull of the sampled sensitivity evaluations, which is then expanded by the error bound M to over-approximate the reachable set of sensitivity values

> *using componentwise* min *and* max *operators, is an over-approximation of the reachable set* $\left\{S_x(t_f; t_0, x_0) \mid x_0 \in \left[\underline{x}, \overline{x}\right]\right\}$ *of first-order sensitivity values.*

The procedure in Proposition 5.3 is illustrated in the two right-most panels of Fig. 5.3 for a uniform grid sampling (left panel). Note that for a 2-dimensional state space, the first-order sensitivity matrix takes values in $\mathbb{R}^{2\times2}$, but for the purpose of visualization the sensitivity is sketched in the 2D plane. We first evaluate numerically the first-order sensitivity matrix $S_x(t_f; t_0, x_0)$ for each initial state x_0 in the sampling set as in the middle of Fig. 5.3. The whole reachable set of the first-order sensitivity $\left\{S_x(t_f; t_0, x_0) \mid x_0 \in \left[\underline{x}, \overline{x}\right]\right\}$ is also represented for comparison, although this set is never actually computed.

Next, on the right-hand side of Fig. 5.3, we take the interval hull of the finite number of sensitivity evaluations $\{S_x(t_f; t_0, x_0) \mid x_0 \in X\}$, and then expand it by the error interval $[-M, M]$ (where M is defined in Proposition 5.3) to obtain the interval which is guaranteed (from Proposition 5.3) to over-approximate the whole reachable set of the first-order sensitivity $\left\{S_x(t_f; t_0, x_0) \mid x_0 \in \left[\underline{x}, \overline{x}\right]\right\}$.

Discussion. By combining a sampling-based strategy with the soundness of interval analysis, the proposed approach presents significant advantages over the methods in Sects. 5.2 and 5.3. The first advantage is that, although both methods in Sect. 5.3 and this section rely on sampling, the interval computed in Proposition 5.3 is a guaranteed over-approximation of the set of first-order sensitivity values, while the one from Sect. 5.3 offers no such guarantee. This is achieved by expanding the sampled evaluations of S_x with the error bound M in Proposition 5.3 instead of solely relying on sampling. On the other hand, the cost of this guarantee is the additional two-step procedure using interval analysis (Lemmas 5.1 and 5.2) and requiring knowledge of the first-order and second-order Jacobian bounds as in Assumption 5.3.

The second advantage is that this approach may result in significantly less conservative over-approximations than the one relying only on interval analysis in Sect. 5.2. In particular, we note that the error bound M in Proposition 5.3 converges to $\mathbf{0}_{n_x \times n_x}$ when the dispersion of the sampling set $d(X)$ goes to zero. This means that if the sampling set is defined as a uniform grid of $[\underline{x}, \overline{x}]$, the interval over-approximation in Proposition 5.3 can be made arbitrarily tight to the set of first-order sensitivity values by increasing the number of samples. This reduced conservativeness, however, comes at the cost of an increased computational complexity due to the higher number of sampled states for which $S_x(t_f; t_0, x_0)$ needs to be numerically evaluated. In summary, the cardinality of the sampling set X can be used to tune the tradeoff between conservativeness and computation time of this approach.

The method presented in this section is similar to the quasi-Monte Carlo approach detailed in Chap. 7. The main difference is that the quasi-Monte Carlo approach in Chap. 7 is applied directly to the system (1.2), while the present section focuses on the dynamics of the first-order sensitivity matrix (5.5), whose reachable set over-approximation from Proposition 5.3 is then used in the main reachability analysis method of Sect. 5.1.

Further Reading

A first version of the reachability analysis presented in Sect. 5.1 was introduced in Xue et al. (2017), in the particular case where the sensitivity matrices are assumed to be sign-stable. The generalization of this approach to any continuous-time system with bounded sensitivity matrices was then provided in Meyer et al. (2018).

To compute bounds on the sensitivity matrix, the first method based on interval analysis as in Sect. 5.2 was introduced in Meyer et al. (2018) and relies on the results from Althoff et al. (2007), Althoff (2010) presented in Sect. 2.3. The creation of the dynamical system describing the sensitivity matrix (5.5) using the chain rule can be found, for example, in Donzé and Male (2007). Instead of including the constant inputs in the state variable as done in Sect. 5.1 to simplify the notation, it is possible to keep the constant inputs in a separate variable. In this case, the sensitivity matrix with respect to the input $S_p(t_f; t_0, x_0, p) = \frac{\partial \Phi(t_f; t_0, x_0, p)}{\partial p}$ can also be bounded with the methods in Sect. 2.3 applied to the affine dynamical system describing its variations and obtained as in Khalil (2001).

The simulation-based approach to approximate the sensitivity bounds in Sect. 5.3 was also proposed in Meyer et al. (2018). More details on the falsification process used in this section can be found, for example, in Kapinski et al. (2016).

The third approach in Sect. 5.4 relying on the second-order sensitivity matrices was introduced in Meyer and Arcak (2020). As for Sect. 5.2, this method mainly relies on interval arithmetics (see Sect. 2.1 or Jaulin et al. (2001)) and the computation of reachable sets and reachable tubes for the continuous-time affine systems describing the dynamics of the first-order and second-order sensitivity matrices. In the introduction of Sect. 5.4, the 2D flattening of the derivative of a functional matrix and the associated matrix product involving the Kronecker product are motivated by the notion of *semi-tensor product* in Cheng et al. (2012). Alternative derivations of second-order sensitivity equations have been obtained in Choi et al. (2016) for differential algebraic equations and in Geng and Hiskens (2019) for hybrid systems. The results related to the notion of dispersion of a gridded sampling are motivated by those on quasi-Monte Carlo methods in Tempo et al. (2012).

Although not detailed in this book, there exists other approaches for reachability analysis relying on sensitivity matrices, which are based on the theory of differential inequalities to obtain time-varying interval bounds using the Mean Value Theorem (Shen and Scott 2018), polyhedral over-approximations using affine and interval relaxations (Harwood and Barton 2018), and non-convex enclosures using Taylor models (Villanueva et al. 2015). The paper Peric et al. (2017) also focuses on bounding sensitivity matrices by polynomial models.

References

Althoff M (2010) Reachability analysis and its application to the safety assessment of autonomous cars. Ph.D. thesis, Technische Universität München

Althoff M, Stursberg O, Buss M (2007) Reachability analysis of linear systems with uncertain parameters and inputs. In: 46th IEEE conference on decision and control. IEEE, pp 726–732

Cheng D, Qi H, Zhao Y (2012) An introduction to semi-tensor product of matrices and its applications. World Scientific

Choi H, Seiler PJ, Dhople SV (2016) Propagating uncertainty in power-system DAE models with semidefinite programming. IEEE Trans Power Syst 32(4):3146–3156

Donzé A, Maler O (2007) Systematic simulation using sensitivity analysis. In: International workshop on hybrid systems: computation and control. Springer, pp 174–189

Geng S, Hiskens IA (2019) Second-order trajectory sensitivity analysis of hybrid systems. IEEE Trans Circuits Syst I Regul Pap 66(5):1922–1934

Harwood SM, Barton PI (2018) Affine relaxations for the solutions of constrained parametric ordinary differential equations. Optim Control Appl Methods 39(2):427–448

Jaulin L, Kieffer M, Didrit O, Walter E (2001) Applied interval analysis: with examples in parameter and state estimation, robust control and robotics, vol 1. Springer Science & Business Media

Kapinski J, Deshmukh JV, Jin X, Ito H, Butts K (2016) Simulation-based approaches for verification of embedded control systems: an overview of traditional and advanced modeling, testing, and verification techniques. IEEE Control Syst 36(6):45–64

Khalil HK (2001) Nonlinear systems, 3rd edn. Pearson

Meyer PJ, Arcak M (2020) Interval reachability analysis using second-order sensitivity. In: Proceedings of the 21st IFAC world congress (virtual), pp 1851–1856

Meyer PJ, Coogan S, Arcak M (2018) Sampled-data reachability analysis using sensitivity and mixed-monotonicity. IEEE Control Syst Lett 2(4):761–766

Peric ND, Villanueva ME, Chachuat B (2017) Sensitivity analysis of uncertain dynamic systems using set-valued integration. SIAM J Sci Comput 39(6):A3014–A3039

Shen K, Scott JK (2018) Mean value form enclosures for nonlinear reachability analysis. In: 2018 IEEE conference on decision and control (CDC). IEEE, pp 7112–7117

Tempo R, Calafiore G, Dabbene F (2012) Randomized algorithms for analysis and control of uncertain systems: with applications. Springer Science & Business Media

Villanueva ME, Houska B, Chachuat B (2015) Unified framework for the propagation of continuous-time enclosures for parametric nonlinear ODEs. J Glob Optim 62(3):575–613

Xue B, Fränzle M, Mosaad PN (2017) Just scratching the surface: partial exploration of initial values in reach-set computation. In: 56th IEEE conference on decision and control, pp 1769–1775

Chapter 6
Growth Bounds

In this chapter, we consider reachability methods that characterize how a set centered around a system trajectory grows or contracts as the trajectory evolves. In particular, we consider *interval growth bounds*[1] which characterize the growth and contraction of intervals. An interval over-approximation of the reachable set from an interval of initial states can be obtained by computing a single successor state from the center of the initial set and bounding the growth of neighboring trajectories.

We first describe the general method for over-approximating reachable sets with interval growth bounds in Sect. 6.1. Then, we describe three methods to compute growth bounds which can be used with the general method. Section 6.2 describes a growth bound derived from a *matrix measure* of the state Jacobian matrix. Section 6.3 describes a growth bound derived from a componentwise upper bound on the state Jacobian matrix. This bound is less conservative than the bound of Sect. 6.2, but requires more computation. Section 6.4 describes a general class of interval growth bounds that can be computed using matrix measures of a blockwise partition of the state Jacobian matrix. This class of bounds allows for a great amount of flexibility in computational efficiency and conservativeness, and can exploit block structure present in the system.

6.1 Interval Reachability with Growth Bounds

In this section, we describe the general method for computing an interval over-approximation of the reachable set using a growth bound. Typically, growth bounds characterize the growth of the system using norm balls, yielding a norm ball over-

---

[1] Such bounds are also known as *contraction metrics* (Aylward et al. 2008), *discrepancy functions* (Fan et al. 2016), and the property of *incremental stability* (Angeli 2002). Despite the name, growth bounds also describe the contraction of sets.

© The Author(s), under exclusive license to Springer Nature Switzerland AG 2021   61
P.-J. Meyer et al., *Interval Reachability Analysis*,
SpringerBriefs in Control, Automation and Robotics,
https://doi.org/10.1007/978-3-030-65110-7_6

approximation for the reachable set. Since intervals are geometrically equivalent to infinity norm balls with a scaling factor along each direction, growth bounds can be used to solve Problem 1.2.

Requirements and Limitations. This method applies to continuous-time systems of the form (1.2), for which an interval growth bound exists. The growth bound is described by a function G with the following properties.

Assumption 6.1 Given an invariant state space $X \subseteq \mathbb{R}^{n_x}$ containing the initial set, there exists a vector-valued function $G : \mathbb{R} \times \mathbb{R} \times \mathbb{R}_+^{n_x} \times \mathbb{R}_+^{n_p} \to \mathbb{R}_+^{n_x}$, called the *interval growth bound function*, for which the inequality

$$\left| \Phi(t_f; t_0, x, \mathbf{p}) - \Phi(t_f; t_0, y, \mathbf{q}) \right| \leq G\left(t_f, t_0, |x - y|, \left| \overline{p} - \underline{p} \right| \right)$$

holds for all $x, y \in X$ and $\mathbf{p}, \mathbf{q} : [t_0, t_f] \to \left[\underline{p}, \overline{p} \right]$, where the absolute value $| \cdot |$ and inequality $\leq$ are applied componentwise.

G is also monotone in the last two arguments: for all $z, w \in \mathbb{R}_+^{n_x}$ and $p, q \in \mathbb{R}_+^{n_p}$ such that $z \geq w$ and $p \geq q$, we have

$$G(t_f, t_0, z, p) \geq G(t_f, t_0, w, q),$$

using componentwise inequalities.

Generally, additional assumptions on the system are required to compute a growth bound function satisfying Assumption 6.1. However, these additional assumptions are different for each of the methods presented in Sects. 6.2–6.4, so they are discussed individually in their respective sections.

Reachability Method. This method computes the reachable set over-approximation in two steps, illustrated in Fig. 6.1. The first step is to compute the successor state $\Phi(t_f; t_0, x^*, p^*)$, where $x^* = \frac{1}{2}(\overline{x} + \underline{x})$ and $p^* = \frac{1}{2}(\overline{p} + \underline{p})$ are the centers of $[\underline{x}, \overline{x}]$ and $\left[\underline{p}, \overline{p} \right]$. The second step is to evaluate the growth bound $G(t_f, t_0, [x], [p])$, where $[x] = \frac{1}{2}(\overline{x} - \underline{x})$ and $[p] = \frac{1}{2}(\overline{p} - \underline{p})$ are the half-widths of $[\underline{x}, \overline{x}]$ and $\left[\underline{p}, \overline{p} \right]$. The successor state and growth bound then form the center and half-width of a reachable set over-approximation.

Proposition 6.1 *Under Assumption 6.1, the interval*

$$\left[\Phi\left(t_f; t_0, x^*, p^*\right) - G\left(t_f, t_0, [x], [p]\right), \Phi\left(t_f; t_0, x^*, p^*\right) + G\left(t_f, t_0, [x], [p]\right) \right]$$

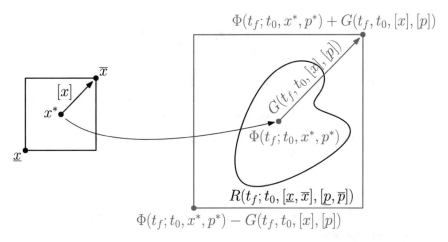

Fig. 6.1 Using a growth bound to over-approximate a reachable set proceeds in two steps: computing a single successor state and evaluating the growth bound. The successor state and growth bound form the center and half-width of the reachable set over-approximation

is an over-approximation of the reachable set $R\left(t_f; t_0, \left[\underline{x}, \overline{x}\right], \left[\underline{p}, \overline{p}\right]\right)$ of (1.2), solving Problem 1.2.

By the monotonicity of G, the interval $\left[-G\left(t_f, t_0, [x], [p]\right), G\left(t_f, t_0, [x], [p]\right)\right]$ over-approximates the growth bound intervals for all pairs in the initial and input sets. This ensures that the interval in Proposition 6.1 over-approximates the amount by which the successor of any point in the initial set can deviate from the successor of the center point, making it an over-approximation of the reachable set.

Discussion. This method can be applied with any interval growth bound. However, the method used to compute each bound affects the efficiency and conservativeness of the overall method, and should, therefore, be chosen carefully. The growth bounds described in the next three sections offer several choices on the tradeoff between computational efficiency and conservativeness. Generally, Sect. 6.2 gives faster and more conservative bounds, Sect. 6.3 gives slower but more accurate bounds, and the family of bounds described in Sect. 6.4 allows for a compromise between the two extremes.

6.2 Growth Bounds from Matrix Measures

In this section, we describe a growth bound derived from an upper bound on the *infinity matrix measure* of the state Jacobian of the system. The infinity measure of a matrix A (with elements A_{ij}) is defined as

$$\mu_\infty(A) = \max_i \left(A_{ii} + \sum_{i \neq j} |A_{ij}| \right). \tag{6.1}$$

Similar to a norm, the infinity measure characterizes the "magnitude" of a matrix. Unlike norms, however, the measure may be negative. This allows the measure to characterize contraction as well as growth.

Requirements and Limitations. This method applies to systems with *additive input*, i.e., whose dynamics are of the form

$$\dot{x} = f(t, x) + p. \tag{6.2}$$

The Jacobian matrix J_x must also satisfy a boundedness condition with respect to the infinity measure.

Assumption 6.2 Given an invariant state space $X \subseteq \mathbb{R}^{n_x}$ containing the initial set, there exists a constant $c \in \mathbb{R}$ such that $\mu_\infty(J_x(t, x, p)) \leq c$ for all $t \in [t_0, t_f]$, $x \in X$, $p \in \left[\underline{p}, \overline{p} \right]$.

If the elements of the state Jacobian are bounded for all t, x, and p, then we can take $X = \mathbb{R}^{n_x}$, and the method can be applied to any initial set and inputs. Otherwise, we need to find another invariant state space over which they can be bounded.

Method. This growth bound is computed as a linear combination of the half-widths of the initial set and input set, with coefficients that depend on the value c in Assumption 6.2.

Proposition 6.2 *Under Assumption 6.2, the function*

$$G(t_f, t_0, [x], [p]) = e^{c(t_f - t_0)}[x] + \int_{t_0}^{t_f} e^{c(t_f - \tau)}[p] d\tau$$

$$= e^{c(t_f - t_0)}[x] + \frac{e^{c(t_f - t_0)} - 1}{c}[p] \tag{6.3}$$

is a growth bound function which satisfies Assumption 6.1 for the system (6.2).

Discussion. Although we can efficiently compute this growth bound, the resulting reachable set over-approximation in Proposition 6.1 is generally very conservative. This is because the growth bound must account for the worst-case growth, and apply it to each dimension: the growth in many directions may be overestimated this way.

Instead of requiring the system to have additive inputs, we can consider general continuous-time dynamics (1.2) by approximating the effect of the input by an additive term. In this case, if there exists a vector $b \in \mathbb{R}^{n_x}$ such that

$$\left| f(t, x, p) - f(t, x, p^*) \right| \leq b \tag{6.4}$$

holds for all $x \in X$, $t \in [t_0, t_f]$, and $p \in \left[\underline{p}, \overline{p} \right]$, then we can simply use b instead of $[p]$ in (6.3).

6.3 Growth Bounds from Componentwise Jacobian Bounds

Instead of bounding the state Jacobian matrix with a single scalar measure to derive a growth bound, we can bound each component of the Jacobian separately. The growth bound derived this way is generally less conservative than the bound derived from the matrix measure, but it is more expensive to compute.

Requirements and Limitations. This growth bound also requires an additive input, so we still assume that the dynamics are of the form $\dot{x} = f(t, x) + p$ as in (6.2). However, the technique described in the discussion of Sect. 6.2 where we replace $[p]$ with the vector b defined in (6.4) can also be applied to this method.

Instead of a single bound on the measure of the whole state Jacobian matrix, we assume a separate bound on each component.

Assumption 6.3 Given an invariant state space $X \subseteq \mathbb{R}^{n_x}$ containing the initial set, there exists a constant matrix $C \in \mathbb{R}^{n_x \times n_x}$ such that

$$\begin{cases} J_{x,ii}(t, x, p) \leq C_{ii} \\ \left| J_{x,ij}(t, x, p) \right| \leq C_{ij}, \quad i \neq j \end{cases}$$

for all $t \in [t_0, t_f]$, $x \in X$, $p \in \left[\underline{p}, \overline{p} \right]$.

Method. This growth bound function is created similarly to the one in Proposition 6.2, but the exponentials need to be replaced by matrix exponentials since C is a matrix.

Proposition 6.3 *Under Assumption 6.3, the function*

$$G(t_f, t_0, [x], [p]) = e^{C(t_f - t_0)}[x] + \int_{t_0}^{t_f} e^{C(t_f - \tau)}[p] d\tau. \tag{6.5}$$

is a growth bound function which satisfies Assumption 6.1 for the system (6.2).

Discussion. The use of matrix exponentials in (6.5) has two consequences. First, it means that (6.5) provides a much less conservative bound than (6.3), since it is not restricted to scaling each element of the half-width by the worst-case growth rate. Second, it means that (6.5) takes longer to compute than (6.3), since computing matrix exponentials is much more difficult than computing scalar exponentials.

However, we do not need to compute the full exponential or its integral. In fact, the quantity in (6.5) is the solution to the affine differential equation $\dot{z} = Cz + [p]$ at time t_f with initial condition $z(t_0) = [x]$. We can, therefore, compute (6.5) by numerically integrating this differential equation. This is faster than computing (6.5) directly, but still slower than computing (6.3).

6.4   Growth Bounds from Blockwise Matrix Measures

In this section, we describe a growth bound derived from a generalization of the matrix measure bound in Sect. 6.2. Instead of bounding the entire state Jacobian matrix using a single measure, this bound uses a partition of the Jacobian matrix $J_x(t, x, p)$ into blocks

$$J_{x,ij}(t, x, p) \in \mathbb{R}^{n_i \times n_j}, \quad i, j = 1, \dots, k,$$

and bounds each block separately. This allows the growth bound to exploit block structure in the system Jacobian, if any is present.

In addition to the infinity matrix measure (6.1), this growth bound also relies on the matrix infinity norm defined as

$$\|A\|_\infty = \max_i \sum_j |A_{ij}|.$$

Requirements and Limitations. As before, we assume that the system has additive input as in (6.2). When this is not the case, we can still replace $[p]$ by the vector b defined in (6.4) as discussed at the end of Sect. 6.2.

Each Jacobian block $J_{x,ij}(t, x, p)$ must also satisfy a boundedness condition with respect to the infinity measure if it is a diagonal block, or with respect to the infinity norm if it is an off-diagonal block.

Assumption 6.4   Given an invariant state space $X \subseteq \mathbb{R}^{n_x}$ containing the initial set, there exists a constant $k \times k$ matrix C such that

$$\begin{cases} \mu_\infty \left(J_{x,ii}(t, x, p) \right) \le C_{ii} \\ \|J_{x,ij}(t, x, p)\|_\infty \le C_{ij}, \quad i \ne j \end{cases}$$

for all $t \in [t_0, t_f]$, $x \in X$, $p \in \left[\underline{p}, \overline{p} \right]$.

Method. To compute this growth bound, we first partition the half-widths of the initial set and input set with the same structure as the Jacobian. We denote these partitions as

$$[x] = \begin{pmatrix} [x]_1 \\ [x]_2 \\ \vdots \\ [x]_k \end{pmatrix}, \quad [p] = \begin{pmatrix} [p]_1 \\ [p]_2 \\ \vdots \\ [p]_k \end{pmatrix},$$

where $[x]_i \in \mathbb{R}^{n_i}$ and $[p]_i \in \mathbb{R}^{n_i}$, $i = 1, \ldots, k$.

As in Proposition 6.3, the blockwise growth bound function defined below uses matrix exponentials. However, these are exponentials of $k \times k$ matrices, whereas Proposition 6.3 uses exponentials of $n_x \times n_x$ matrices.

Proposition 6.4 *Let the vector $r \in \mathbb{R}^k$ be defined as*

$$r = e^{C(t_f - t_0)} \begin{pmatrix} \|[x]_1\|_\infty \\ \|[x]_2\|_\infty \\ \vdots \\ \|[x]_k\|_\infty \end{pmatrix} + \int_{t_0}^{t_f} e^{C(t_f - \tau)} \begin{pmatrix} \|[p]_1\|_\infty \\ \|[p]_2\|_\infty \\ \vdots \\ \|[p]_k\|_\infty \end{pmatrix} d\tau.$$

Then, under Assumption 6.4, the function

$$G(t_f, t_0, [x], [p]) = \begin{pmatrix} r_1 \mathbf{1}_{n_1} \\ \vdots \\ r_k \mathbf{1}_{n_k} \end{pmatrix}$$

is a growth bound function which satisfies Assumption 6.1 for the system (6.2).

Discussion. Since each block decomposition leads to a different bound, this section defines an entire family of growth bounds. In fact, the bounds described in Sects. 6.2 and 6.3 are both special cases of the blockwise bound for specific block structures. The measure-based bound of Sect. 6.2 corresponds to the decomposition with a single block, $J_{x,11}(t, x, p) = J_x(t, x, p)$. At the other extreme, the componentwise bound of Sect. 6.3 sets each block $J_{x,ij}(t, x, p)$ to be the ijth element of $J_x(t, x, p)$.

There are certain systems for which a blockwise bound can make a less conservative over-approximation than the componentwise bound. This is the case when a system has a structure that the blockwise bound can exploit. For instance, consider the system

$$\dot{x} = f(t, x, x_p)$$
$$\dot{x}_p = \mathbf{0}_{n_p},$$

where the *states* x_p are used to represent constant parameter uncertainties. For this system, using the componentwise bound could lead to a more conservative estimate than a blockwise bound with a structure that puts all of the x_p states into a single block.

Further Reading

Most reachability methods based on growth bounds, such as those presented in Kapela and Zgliczyński (2009), Maidens and Arcak (2014), Rungger and Zamani (2016), Reissig et al. (2016), Arcak and Maidens (2018), follow the general form described in Sect. 6.1. Growth bound methods are also referred to by several other names, such as *contraction bounds*, *contraction metrics* (Aylward et al. 2008), and *discrepancy functions* (Fan et al. 2016). Growth bounds can characterize both upper bounds on the growth between trajectories and lower bounds on the contraction between trajectories: this is why the terms "growth" and "contraction" are both used to describe the same bound.

Growth bounds are a generalization of the property of *incremental stability* (Angeli 2002), which states that arbitrary trajectories contract towards each other over time. The study of incremental stability properties is also called *contraction theory*: (Aminzare and Sontag 2014) provides a summary of general contraction-theoretic methods in control theory.

The measure-based growth bound in Sect. 6.2 is based on a reachability method presented in Maidens and Arcak (2014). This method uses a general matrix measure to compute reachable set over-approximations with norm ball geometry, with the specific norm ball depending on the measure. Section 6.2 considers the specific case of infinity measures because this measure yields interval over-approximations defined as infinity norm balls. The method in Maidens and Arcak (2014) also considers weighted matrix measures, which can allow for tighter over-approximations.

The componentwise growth bound in Sect. 6.3 is based on a reachability method presented in Reissig et al. (2016). This method considers an arbitrary componentwise bound for general systems, and provides Proposition 6.3 as a special case for continuously differentiable systems.

The blockwise growth bound in Sect. 6.4 is based on an algorithm presented in Kapela and Zgliczyński (2009). This growth bound may also employ other measures than the infinity measure. Each block of the partition can even use a different matrix measure, which can be optimized to reduce the conservativeness of the over-approximation (Arcak and Maidens 2018).

Although this chapter focuses on growth bounds based on matrix measures, they can also be computed by other means. For instance, growth bounds which provide locally optimal approximations to the reachable set may be computed from the system dynamics (Fan et al. 2016).

References

Aminzare Z, Sontag ED (2014) Contraction methods for nonlinear systems: a brief introduction and some open problems. In: 53rd IEEE conference on decision and control. IEEE, pp 3835–3847

Angeli D (2002) A Lyapunov approach to incremental stability properties. IEEE Trans Autom Control 47(3):410–421

Arcak M, Maidens J (2018) Simulation-based reachability analysis for nonlinear systems using componentwise contraction properties. In: Principles of modeling. Springer, pp 61–76

Aylward EM, Parrilo PA, Slotine JJE (2008) Stability and robustness analysis of nonlinear systems via contraction metrics and SOS programming. Automatica 44(8):2163–2170

Fan C, Kapinski J, Jin X, Mitra S (2016) Locally optimal reach set over-approximation for nonlinear systems. In: 2016 international conference on embedded software (EMSOFT). IEEE, pp 1–10

Kapela T, Zgliczyński P (2009) A Lohner-type algorithm for control systems and ordinary differential inclusions. Discrete Contin Dyn Syst Ser B 11(2):365–385

Maidens J, Arcak M (2014) Reachability analysis of nonlinear systems using matrix measures. IEEE Trans Autom Control 60(1):265–270

Reissig G, Weber A, Rungger M (2016) Feedback refinement relations for the synthesis of symbolic controllers. IEEE Trans Autom Control 62(4):1781–1796

Rungger M, Zamani M (2016) SCOTS: A tool for the synthesis of symbolic controllers. In: Proceedings of the 19th international conference on hybrid systems: computation and control, pp 99–104

Chapter 7
Sampling-Based Methods

This chapter considers methods that compute reachable set approximations by evaluating the successor function on a finite set of sample points. Such *sampling-based*
methods can work under mild system assumptions, but they require the successor
function to be evaluated once for each sample point. Sampling-based methods are,
therefore, broadly applicable and also computationally expensive.

This chapter presents two sampling-based methods. The method presented in
Sect. 7.1 constructs a grid over the initial states and inputs, and evaluates the successor
state of each grid point. Due to the grid-based sampling scheme, the number of sample
points used by this *quasi-Monte Carlo* method increases exponentially with the
state and input dimensions. To avoid the exponential scaling, the method presented
in Sect. 7.2 uses a randomized sampling scheme. Due to its probabilistic nature,
this *Monte Carlo* method is not guaranteed to produce an over-approximation of
the reachable set. However, it can produce an approximation which is close to the
reachable set in a probabilistic sense, and the number of samples needed to do so
increases only linearly in the state dimension.

Sampling-based methods can be applied to both continuous-time and discrete-
time systems. The methods are presented for the continuous-time case in this chapter:
the discrete-time case is identical, except that we seek to solve Problem 1.1 instead
of Problem 1.2, and use the discrete-time successor function $F(t, x, p)$ instead of
the continuous-time successor function $\Phi(t_f; t_0, x_0, \mathbf{p})$.

7.1 Quasi-Monte Carlo

The method presented in this section uses a deterministic scheme to select the sample
points it computes. Sampling-based methods that select sample points determinis-
tically are called quasi-Monte Carlo methods, in contrast to Monte Carlo methods,
which use randomly selected samples. The samples are computed by constructing

© The Author(s), under exclusive license to Springer Nature Switzerland AG 2021   71
P.-J. Meyer et al., *Interval Reachability Analysis*,
SpringerBriefs in Control, Automation and Robotics,
https://doi.org/10.1007/978-3-030-65110-7_7

uniform grids over the initial set and input set and evaluating the successor function for all pairs of grid points. This set of samples, augmented by an error bound, can be used to solve Problem 1.2.

Requirements and Limitations. For this method, we consider only constant inputs in the range $\left[\underline{p}, \overline{p}\right]$. In order to compute an over-approximation of the reachable set from the grid of sample points, we need to know a function bounding the distance between two successor states based on their initial states and inputs. We express this bound with the function K in the following assumption.

Assumption 7.1 Given an invariant state space $X \subseteq \mathbb{R}^{n_x}$ containing the initial set, there exists a function $K : \mathbb{R} \times \mathbb{R} \times \mathbb{R}_+ \times \mathbb{R}_+ \to \mathbb{R}_+$ such that the inequality

$$\left\| \Phi(t_f; t_0, x, p) - \Phi(t_f; t_0, y, q) \right\|_\infty \leq K\left(t_f, t_0, \|x - y\|_\infty, \|p - q\|_\infty\right)$$

holds for all $x, y \in X$ and constant inputs $p, q \in \left[\underline{p}, \overline{p}\right]$.

K is also monotone in the last two arguments: for all $a, b, c, d \in \mathbb{R}_+$ such that $a \geq b$ and $c \geq d$, we have

$$K(t_f, t_0, a, c) \geq K(t_f, t_0, b, d).$$

Although the function K in Assumption 7.1 is similar to the growth bound function G in Assumption 6.1, the main difference is that K defines a scalar bound based on scalar arguments (thus using the infinity norm to define distances between initial states, inputs, and successor states), while G gives a vector bound and uses vector arguments (hence defining distances with the componentwise absolute value). For a continuous-time system with additive input as in (6.2) and satisfying Assumption 6.2, a function K satisfying Assumption 7.1 can then be defined as the scalar version of the matrix measure-based growth bound from Proposition 6.2

$$K\left(t_f, t_0, \|x - y\|_\infty, \|p - q\|_\infty\right) = e^{c(t_f - t_0)} \|x - y\|_\infty + \frac{e^{c(t_f - t_0)} - 1}{c} \|p - q\|_\infty,$$

where c is the constant defined in Assumption 6.2.

Another case where Assumption 7.1 holds is when the system is Lipschitz continuous with a Lipschitz constant L that holds for both the initial state and the input. In this case, the bound is

$$K\left(t_f, t_0, \|x - y\|_\infty, \|p - q\|_\infty\right) = e^{L(t_f - t_0)} \left(\|x - y\|_\infty + \|p - q\|_\infty\right).$$

This bound is generally worse than the one derived from the matrix measure: the Lipschitz constant L is always positive, while the matrix measure c may be negative.

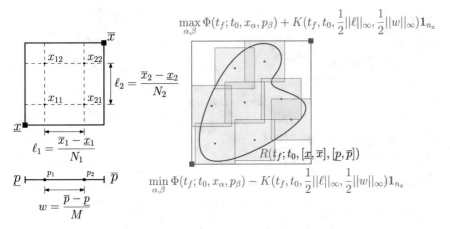

Fig. 7.1 *Left*: the grids of x_α and p_β sampling the initial set and input set, shown here for a two-dimensional state space and one-dimensional input, with $N_1 = N_2 = M = 2$. *Right*: the interval over-approximation of the reachable set derived from the successor states of the $N_1 N_2 M = 8$ sample points and K. Each sample is shown surrounded by an interval of half-width $K\left(t_f, t_0, \frac{1}{2}\|\ell\|_\infty, \frac{1}{2}\|w\|_\infty\right) \mathbf{1}_{n_x}$

Reachability Method. The first step is to construct the grids over the intervals of initial states and inputs, as illustrated on the left side of Fig. 7.1. For each state dimension i, we select a number N_i of grid points to use along that dimension. Similarly, for each input dimension j, we select a number M_j of grid points to use along that dimension. The distance between grid points along each state dimension is then $\ell_i = \left(\overline{x}_i - \underline{x}_i\right)/N_i$, which we collect in the vector ℓ. The distance between grid points along each input dimension is $w_j = \left(\overline{p}_j - \underline{p}_j\right)/M_j$, which we collect in the vector w. The initial state grid is indexed with $\alpha = (\alpha_1, \ldots, \alpha_{n_x})$, $\alpha_i \in \{1, \ldots, N_i\}$, and the input grid with $\beta = (\beta_1, \ldots, \beta_{n_p})$, $\beta_j \in \{1, \ldots, M_i\}$, so that

$$x_\alpha = \begin{pmatrix} \underline{x}_1 + \left(\alpha_1 - \frac{1}{2}\right)\ell_1 \\ \vdots \\ \underline{x}_{n_x} + \left(\alpha_{n_x} - \frac{1}{2}\right)\ell_{n_x} \end{pmatrix}, \qquad p_\beta = \begin{pmatrix} \underline{p}_1 + \left(\beta_1 - \frac{1}{2}\right)w_1 \\ \vdots \\ \underline{p}_{n_p} + \left(\beta_{n_p} - \frac{1}{2}\right)w_{n_p} \end{pmatrix}$$

are the points on the initial state and input grids, respectively.

The second step is to evaluate the successor states $\Phi(t_f; t_0, x_\alpha, p_\beta)$ for each pair (x_α, p_β) of initial state and input on the grids. Since there are $\prod_{i=1}^{n_x} N_i$ points on the initial state grid, and $\prod_{j=1}^{n_p} M_j$ points on the input grid, there are $\prod_{i=1}^{n_x} N_i \prod_{j=1}^{n_p} M_j$ samples to evaluate.

Proposition 7.1 *Under Assumption 7.1, the interval*

$$\left[\min_{\alpha,\beta} \Phi(t_f; t_0, x_\alpha, p_\beta) - K\left(t_f, t_0, \tfrac{1}{2}\|\ell\|_\infty, \tfrac{1}{2}\|w\|_\infty\right) \mathbf{1}_{n_x}, \right.$$

$$\left. \max_{\alpha,\beta} \Phi(t_f; t_0, x_\alpha, p_\beta) + K\left(t_f, t_0, \tfrac{1}{2}\|\ell\|_\infty, \tfrac{1}{2}\|w\|_\infty\right) \mathbf{1}_{n_x} \right],$$

where min *and* max *denote componentwise minimum and maximum, respectively, is an over-approximation of the reachable set* $R\left(t_f; t_0, [\underline{x}, \overline{x}], [\underline{p}, \overline{p}]\right)$ *of (1.2), solving Problem 1.2.*

The over-approximation of Proposition 7.1 is equivalent to taking the interval hull of the set of intervals centered on each successor state $\Phi(t_f; t_0, x_\alpha, p_\beta)$ with half-widths $K\left(t_f, t_0, \tfrac{1}{2}\|\ell\|_\infty, \tfrac{1}{2}\|w\|_\infty\right) \mathbf{1}_{n_x}$, as illustrated on the right side of Fig. 7.1.[1] These intervals act as error bounds on the successor sample points. By Assumption 7.1, the intervals cover the reachable set, ensuring that their interval hull is a solution to Problem 1.2.

Discussion. Although the quasi-Monte Carlo sampling method is presented here as a method to over-approximate reachable sets directly from the successor function, it may also be used as a step in other reachability methods where a set must be over-approximated using samples. For instance, the method of Sect. 5.4 uses quasi-Monte Carlo sampling to bound the sensitivity matrix.

Since a growth bound can be used to satisfy Assumption 7.1, the quasi-Monte Carlo method presented in this section is related to the reachability methods discussed in Chap. 6. The quasi-Monte Carlo method may be viewed as a "divide and conquer" approach to the growth bound methods. Instead of providing a single growth bound for the entire reachable set, the initial set is divided into a grid of cells, and a separate growth bound is established for each cell. The final reachable set over-approximation is the interval hull of the over-approximations computed for each cell, just as shown in Fig. 7.1. This means that the quasi-Monte Carlo method produces a less conservative reachable set over-approximation than a single growth bound, but it does so at a greater computational cost.

The number of samples used by a sampling-based method is called its *sample complexity*. The sample complexity of the quasi-Monte Carlo method scales poorly with the dimensions of the system. In the case where the numbers of samples per dimension N_i and M_j all have the same value d, the total number of samples is $d^{n_x+n_p}$. The sample complexity, therefore, scales *exponentially* with n_x and n_p when the number of grid points along each dimension is fixed.

In general, it is not possible to improve the sample complexity of the quasi-Monte Carlo method by choosing a different sampling scheme without increasing the conservativeness of the over-approximation. This is due to the result known as the *Sukharev inequality*, which places a lower bound on the *dispersion*

---

[1] While equivalent, the formulation of Proposition 7.1 is more efficient to compute, since it avoids having to construct the intervals around each sample point.

$$d_I \left(x^{(1)}, \ldots, x^{(k)} \right) = \sup_{x \in I} \min_{1 \leq i \leq k} \left\| x - x^{(i)} \right\|_\infty$$

of a set of points $\{x^{(i)}\}_{i=1}^k$ in an interval set $I \subset \mathbb{R}^{n_x}$ (Sukharev 1971). Since the approximation error bound $K \left(t_f, t_0, \frac{1}{2}\|\ell\|_\infty, \frac{1}{2}\|w\|_\infty \right)$ is determined by the infinity norm of the distance between sample points, the dispersion of a sample set determines the tightness of the over-approximation, with a lower dispersion being better. According to the Sukharev inequality, the minimum dispersion is attained when the points $x^{(i)}$ are arranged in a regular grid. In order to attain as low a dispersion as the regular grid, any other sampling scheme must use at least as many points, meaning that the sample complexity of that scheme cannot be better than the grid-based scheme.

7.2 Monte Carlo

Since the number of samples used by the quasi-Monte Carlo method increases exponentially with n_x and n_p, its practical use is limited to low-dimensional systems. Additionally, the requirement of Assumption 7.1 limits its generality. As an alternative, this section presents a *Monte Carlo* sampling-based method which controls the sample complexity with a randomized sampling scheme, and which does not rely on Assumption 7.1.

Unlike the other methods presented in this book, the Monte Carlo method cannot solve Problems 1.1 or 1.2, since it is not guaranteed to compute an over-approximation of the reachable set. However, the randomized sampling scheme ensures that the reachable set approximation it computes is highly likely (in other words, with high *confidence*) to be accurate in a probabilistic sense. Specifically, it ensures, with probability greater than $1 - \delta$, that the probability that a pair of initial state and input selected at random yields a successor lying in the reachable set approximation is at least $1 - \epsilon$, where $\epsilon, \delta \in (0, 1)$ are chosen by the user. The probability $1 - \epsilon$ corresponds to the probabilistic accuracy, and the probability $1 - \delta$ is the confidence that the approximation attains this accuracy. The number of samples needed to ensure this probabilistic accuracy and confidence increases only linearly with the state dimension n_x, and does not depend on the input dimension n_p.

Requirements and Limitations. Like the quasi-Monte Carlo method, we consider only constant inputs in the range $\left[\underline{p}, \overline{p} \right]$. The only other requirement is that the successor function should be well-defined and finite for each initial state in the initial set and each input in the input set. No additional system information outside of the successor function, e.g., an expression or bound for the Jacobian, is required.

Reachability Method. The first step is to determine the number of samples N. For this, the user can choose desired values for the parameters $\epsilon, \delta \in (0, 1)$, and then define N as

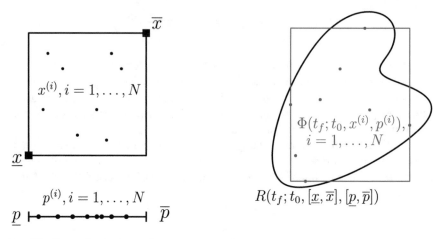

Fig. 7.2 The randomized sampling scheme used by the Monte Carlo method, shown here for a two-dimensional state space and one-dimensional input. *Left*: the sample points $x^{(i)}$ and $p^{(i)}$ are drawn uniformly from the initial set and input set. *Right*: the reachable set approximation computed by the Monte Carlo method is the interval hull of the successor states of the sample points

$$N = \left\lceil \frac{1}{\epsilon} \left(\frac{e}{e-1} \right) \left(\log \frac{1}{\delta} + 2n_x \right) \right\rceil, \tag{7.1}$$

where e is the Euler number and $\lceil \cdot \rceil$ is the *ceiling operator*, i.e., $\lceil x \rceil$ is the smallest integer greater than or equal to x.

The second step is to select the sample points as shown on the left side of Fig. 7.2: for this, we select N points $x^{(i)}$, $i = 1, \ldots, N$ uniformly at random from $[\underline{x}, \overline{x}]$ and N points $p^{(i)}$, $i = 1, \ldots, N$ uniformly at random from $[\underline{p}, \overline{p}]$. The third step is to compute the successor states $\Phi(t_f; t_0, x^{(i)}, p^{(i)})$ for $i = 1, \ldots, N$. The reachable set approximation is then the interval hull of the successor states, as shown on the right side of Fig. 7.2, whose probabilistic accuracy and confidence is formally stated by the following proposition.

Proposition 7.2 *Let P_S be the probability measure corresponding to the joint distribution of N independent samples drawn uniformly from both $[\underline{x}, \overline{x}]$ and $[\underline{p}, \overline{p}]$, and let P_Z be the probability measure corresponding to the distribution of the random variable $Z = \Phi(t_f; t_0, X, P)$, where X is the uniform random variable over $[\underline{x}, \overline{x}]$ and P is the uniform random variable over constant inputs with values in $[\underline{p}, \overline{p}]$. Then the interval hull of the successors of the sample pairs, denoted as*

$$H\left(\left\{\Phi\left(t_f; t_0, x^{(i)}, p^{(i)}\right) \mid i = 1, \ldots, N\right\}\right),$$

satisfies

$$P_S\left(P_Z\left(H\left(\left\{\Phi\left(t_f; t_0, x^{(i)}, p^{(i)}\right) \mid i = 1, \ldots, N\right\}\right)\right) \geq 1 - \epsilon\right) \geq 1 - \delta.$$
(7.2)

The inequality in Proposition 7.2 asserts that the Monte Carlo method can compute reachable set approximations that are accurate, in a probabilistic sense, with high confidence. The accuracy is characterized by the quantity $P_Z\left(H\left(\left\{\Phi\left(t_f; t_0, x^{(i)}, p^{(i)}\right) \mid i = 1, \ldots, N\right\}\right)\right)$. This is the probability that a random sample $z = \Phi(t_f; t_0, x, p)$, where x and p are sampled uniformly from $[\underline{x}, \overline{x}]$ and $[\underline{p}, \overline{p}]$, respectively, lies in the interval hull of $\left\{\Phi\left(t_f; t_0, x^{(i)}, p^{(i)}\right) \mid i = 1, \ldots, N\right\}$. If this quantity is close to 1, then the interval hull contains a large measure, in probability, of the reachable set. The confidence is characterized by the probability that, once a set of sample points is computed, $P_Z\left(H\left(\left\{\Phi\left(t_f; t_0, x^{(i)}, p^{(i)}\right) \mid i = 1, \ldots, N\right\}\right)\right)$ is at least $1 - \epsilon$. Inequality (7.2) asserts that, with probability greater than $1 - \delta$, $P_Z\left(H\left(\left\{\Phi\left(t_f; t_0, x^{(i)}, p^{(i)}\right) \mid i = 1, \ldots, N\right\}\right)\right)$ is at least $1 - \epsilon$. In other words, we have a $1 - \delta$ confidence that the reachable set approximation has accuracy $1 - \epsilon$.

Discussion. The sample complexity of the Monte Carlo method is determined by the number of sample points N in (7.1). Since n_x only appears as a linear term in (7.1), the sample complexity of the Monte Carlo method is linear in the state dimension. The sample complexity is also not affected by the dimension of the input. This is a significant improvement on the complexity of the quasi-Monte Carlo method from Sect. 7.1, which is exponential in both the state and input dimensions. The sample complexity is also affected by the choice of ϵ and δ, with ϵ entering as a $1/\epsilon$ factor and δ as a $\log(1/\delta)$ term. Since the sample complexity increases logarithmically in $1/\delta$, the parameter δ can be set to very low values (to obtain a high confidence $1 - \delta$) with only a small impact on the required number of samples.

As presented, the Monte Carlo method uses uniform sampling distributions to select the $x^{(i)}$ and $p^{(i)}$. However, other sampling distributions may be used. Specifically, let p_X and p_P be probability distributions such that $p_X(x) \geq 0$ if and only if $x \in [\underline{x}, \overline{x}]$, and $p_P(p) \geq 0$ if and only if $p \in [\underline{p}, \overline{p}]$. Then the samples $x^{(i)}$ and $p^{(i)}$ may be drawn from the distributions p_X and p_P, respectively. However, the probabilities in (7.2) must be interpreted with respect to these distributions instead of the uniform distributions.

Instead of requiring constant inputs, we can consider time-varying inputs by defining a probability distribution over input signals. For instance, if the input set is the span of a finite number of basis functions $\mathbf{g}_i$ (i.e., $\mathbf{p}(t) = \sum_i \gamma_i \mathbf{g}_i(t)$ with $\gamma_i \in [\underline{\gamma}_i, \overline{\gamma}_i]$), then we can induce a distribution over the inputs by choosing a distribution over the γ_i, e.g., the uniform distribution. Using time-varying inputs instead of constant inputs does not affect the sample complexity.

Further Reading

The quasi-Monte Carlo sampling method is based on a global optimization technique of the same name. Quasi-Monte Carlo optimization methods evaluate samples of the objective function using a deterministic sampling scheme, and apply a bound on how much the objective can vary between sample points to establish bounds on the extrema (Niederreiter 1992). Quasi-Monte Carlo reachability method treats Problems 1.1 and 1.2 as optimization problems to find the extrema of the components of the successor function.

While the Sukharev inequality implies that the grid-based scheme is optimal for a fixed number of sample points, other sampling schemes are useful when the number of samples is not fixed. For example, a sampling scheme based on a hierarchical refinement of grids is considered in Donzé and Maler (2007) for a sampling-based method which uses as many samples as required to meet an accuracy criterion.

The Monte Carlo method presented in this chapter is an example of a *randomized algorithm*. Randomized algorithms have a wide range of applications to control-theoretic problems. An overview of available methods and some of their applications is available in Tempo et al. (2012).

The probability inequality in Proposition 7.2 is derived using *scenario optimization*. Scenario optimization is a method to approximate the solution to chance-constrained optimization problems which is guaranteed to produce feasible solutions with high probability (Calafiore and Campi 2006). By framing probabilistic analogs of Problems 1.1 and 1.2 as chance-constrained optimization problems, the scenario approach leads to the Monte Carlo method and to Proposition 7.2 (Devonport and Arcak 2020).

References

Calafiore GC, Campi MC (2006) The scenario approach to robust control design. IEEE Trans Autom Control 51(5):742–753

Devonport A, Arcak M (2020) Estimating reachable sets with scenario optimization. In: Learning for dynamics and control

Donzé A, Maler O (2007) Systematic simulation using sensitivity analysis. In: International workshop on hybrid systems: computation and control. Springer, pp 174–189

Niederreiter H (1992) Random number generation and quasi-Monte Carlo methods, vol 63. Siam

Sukharev AG (1971) Optimal strategies of the search for an extremum. USSR Comput Math Math Phys 11(4):119–137

Tempo R, Calafiore G, Dabbene F (2012) Randomized algorithms for analysis and control of uncertain systems: with applications. Springer Science & Business Media

Part II
Applications

Chapter 8
Safety and Reachability Verification

In this chapter, we apply some of the interval reachability analysis methods described in this book to the problem of verifying reachability and safety specifications. A reachability specification considers a *target set* $T \subseteq \mathbb{R}^{n_x}$, a time range $[t_0, t_f]$, and a set of admissible initial states and inputs, and is satisfied if the reachable set is contained in T under these conditions. Safety specifications consider a possibly time-varying *unsafe set* $U(t) \subseteq \mathbb{R}^{n_x}$, $t \in [t_0, t_f]$. The safety specification is satisfied if $\Phi(t; t_0, x_0, \mathbf{p}) \notin U(t)$ for all admissible initial states, inputs, and times in the range $[t_0, t_f]$.

A reachability specification can be verified by computing a reachable set over-approximation: if it is contained entirely inside the target set, then the specification is satisfied. To evaluate the safety specification, we compute an approximation to the reachable tube for the time range $[t_0, t_f]$ by computing reachable set over-approximations at a set of sample times. If the approximation of the reachable tube does not intersect with $U(t)$, then the safety specification is satisfied.

We use several of the methods in this book to compute the over-approximations and to compare their performance in terms of tightness and computation time. In Sect. 8.1, we consider a safety specification for a tunnel diode oscillator which assures that the current through a sensitive component remains below a safe level. In Sect. 8.2, we consider both a safety and a reachability specification for a quadrotor model, which assert that a certain control policy safely brings the quadrotor to a specified height while avoiding large fluctuations.

8.1   Tunnel Diode Oscillator

System Description. The tunnel diode oscillator system is a circuit model used as a verification benchmark in Frehse et al. (2006). The circuit is shown on the left side of Fig. 8.1. The tunnel diode $\displaystyle\bigtriangledown$ has a non-monotone voltage-current relationship, shown on the right side of Fig. 8.1, which exhibits negative resistance for $v_D \in [0.055\,\text{V}, 0.35\,\text{V}]$. This behavior causes the circuit to oscillate with a stable limit cycle. The state equations for this system, expressed in terms of the tunnel diode voltage and the inductor current, are

$$\frac{dv_D}{dt} = \frac{1}{C}\left(-i_D(v_D) + i_L\right)$$
$$\frac{di_L}{dt} = \frac{1}{L}\left(-v_D - Ri_L + V_{in}\right),$$

where $C = 1\,\text{pF}$, $L = 1\,\mu\text{H}$, $R = 200\,\Omega$, and $V_{in} = 0.3\,\text{V}$. We approximate $i_D(v_D)$, the voltage-current relationship of the tunnel diode, with the fifth-degree polynomial that interpolates the points marked on the graph on the right side of Fig. 8.1 and whose second derivative vanishes at $v_D = 0.35\,\text{V}$. This results in a system with $n_x = 2$ states and $n_p = 0$ inputs. The state Jacobian matrix for this system is

$$J_x(t, v_D, i_L) = \begin{pmatrix} -\frac{1}{C}\frac{\partial i_D(v_D)}{\partial v_D} & \frac{1}{C} \\ -\frac{1}{L} & -\frac{R}{L} \end{pmatrix}.$$

Three of the four elements of $J_x(t, v_D, i_L)$ are constant. The only variable element, $-\frac{1}{C}\frac{\partial i_D(v_D)}{\partial v_D}$, can be bounded above, as the polynomial approximation of $-\frac{\partial i_D(v_D)}{\partial v_D}$ attains a global maximum of approximately 0.0074. This means that $J_x(t, v_D, i_L)$ can be made sign-stable with the shifting matrix

$$L_x = \begin{pmatrix} -\frac{0.0074}{C} & 0 \\ 0 & 0 \end{pmatrix},$$

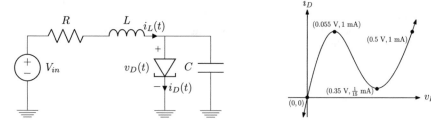

Fig. 8.1 *(Left)*: circuit model for the tunnel diode oscillator. *(Right)*: graph of the current i_D flowing through the tunnel diode as a function of the voltage v_D across it

which shows that the system satisfies Assumption 4.1. Since each diagonal element of $J_x(t, v_D, i_L)$ attains a finite upper bound, and the magnitude of each off-diagonal element is bounded, the system satisfies Assumption 6.3 using the matrix

$$\begin{pmatrix} \frac{0.0074}{C} & \frac{1}{C_R} \\ \frac{1}{L} & -\frac{R}{L} \end{pmatrix},$$

as the componentwise Jacobian bound.

Verification Problem. The safety specification for this system imposes a maximum admissible current through the inductor. Small inductors, particularly at the integrated circuit scale, only admit small currents: beyond a certain threshold, large currents can cause permanent damage. The specification ensures that the inductor current remains below 1.2 mA over the course of a single oscillation, which takes approximately 15 ns. This corresponds to an unsafe set $U = \{(v_D, i_L)|i_L \geq 1.2\,\text{mA}\}$, which must be avoided over the time range $[t_0, t_f] = [0, 15\,\text{ns}]$. We take the set of initial states to be the interval $v_L(0) \in [0.45\,\text{V}, 0.5\,\text{V}]$, $i_L(0) = 0.6\,\text{mA}$. To compute the reachable tube approximation, we compute reachable set over-approximations at 500 evenly spaced points in the range $[t_0, t_f]$.

Figure 8.2 shows the reachable tube approximations computed using the continuous-time mixed monotonicity from Sect. 4.1, sampled-data mixed monotonicity from Sect. 5.1, growth bound from Sects. 6.1 and 6.3, and probabilistic Monte Carlo methods from Sect. 7.2. For the sampled-data mixed-monotonicity method, the bounds on the sensitivity matrix are computed with the sampling technique of Sect. 5.3 using 500 samples. The Monte Carlo method used probabilistic parameters $\epsilon = 0.05$, $\delta = 10^{-9}$. On a computer with two physical cores at 2.6 GHz running MATLAB, computation took 23 s for continuous-time mixed monotonicity, 220 s for sampled-data mixed monotonicity, 19 s for growth bound, and 25 s for Monte Carlo. Three of the four reachability methods verified that the safety specification holds

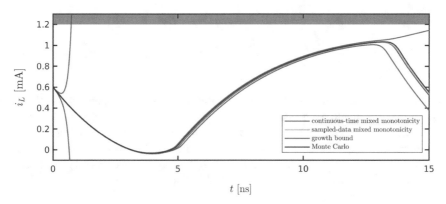

Fig. 8.2 Reachable tube approximations for the inductor current i_L in the tunnel diode oscillator system. To satisfy the safety specification, i_L must not enter the unsafe region (red)

over the course of one period. The growth bound method gives an approximation that is much more conservative than that of the others for this system due to the large off-diagonal terms in the contraction matrix (recall that $1/C = 10^{12}\,\mathrm{F}^{-1}$). With such large terms, the matrix exponential in the growth bound attains large values even when t is on the order of $10^{-9}\,\mathrm{s}$.

8.2   Quadrotor Model

System Description. We consider a quadrotor model based on a safety verification benchmark problem in ARCH-COMP (Immler et al. 2018), a friendly competition between formal verification tools held at the *Applied Verification for Continuous and Hybrid Systems* (ARCH) workshop. The quadrotor model used in the benchmark is a system with 12 states and 4 inputs, whose dynamics are derived from Newton's equations for a rigid body with 3 translational and 3 rotational degrees of freedom. However, the state Jacobian matrix for this system contains unbounded terms, which prevents us from finding an invariant of the state space where the Jacobian matrices are bounded, as required in Assumption 4.1 for continuous-time mixed monotonicity and Assumption 6.3 for the growth bound method. To avoid this difficulty, we consider a version of the model with simplified dynamics,[1] given by

$$
\begin{aligned}
\ddot{p}_n &= \frac{F}{m}(-\cos(\phi)\sin(\theta)\cos(\psi) - \sin(\phi)\sin(\psi)) \\
\ddot{p}_e &= \frac{F}{m}(-\cos(\phi)\sin(\theta)\sin(\psi) + \sin(\phi)\cos(\psi)) \\
\ddot{h} &= \frac{F}{m}\cos(\phi)\cos(\theta) - g \\
\ddot{\phi} &= \frac{1}{I_x}\tau_\phi \\
\ddot{\theta} &= \frac{1}{I_y}\tau_\theta \\
\ddot{\psi} &= \frac{1}{I_z}\tau_\psi
\end{aligned}
\tag{8.1}
$$

where p_n, p_e, and h denote the y-axis ("north"), x-axis ("east") and z-axis ("up") position of the quadrotor, and ϕ, θ, and ψ its pitch, roll, and yaw angles respectively, F is the upward force and τ_ϕ, τ_θ, and τ_ψ the torques exerted by the motors, m is the mass of the quadrotor, and I_x, I_y, and I_z are the moments of inertia about the x-, y- and z-axes. Since the Jacobian matrix $J_x(t, x)$ of the system (8.1) is bounded for all $x \in \mathbb{R}^{n_x}$, this system satisfies both Assumptions 4.1 and 6.3 required for

---

[1]Specifically, the dynamics in (8.1) are obtained from the original dynamics by applying a small-angle approximation and neglecting Coriolis force terms.

the application of the continuous-time mixed monotonicity and the componentwise growth bound methods, respectively.

Verification Problem. The ARCH-COMP safety benchmark for this system is to verify the safety of a proportional-derivative (PD) control policy given by

$$F = mg - 10(h - h_{ref}) + 3\dot{h}, \qquad \tau_\phi = -\phi - \dot{\phi}, \qquad \tau_\theta = -\theta - \dot{\theta}, \qquad \tau_\psi = 0,$$
(8.2)

which seeks to guide the quadrotor to an upright position ($\phi = \theta = 0$) at a height of $h_{ref} = 1$ m. In order to maintain a bounded state Jacobian, we also restrict F to the interval $[F_{min}, F_{max}] = \left[\frac{1}{2}mg, \frac{3}{2}mg\right]$. With the control policy (8.2) in place, the system has $n_x = 12$ states and $n_p = 0$ inputs. This problem has both a safety specification and a reachability specification, corresponding to the unsafe set

$$U(t) = \begin{cases} \{x | h \geq 1.4\,\text{m}\}, & t < 1\,\text{s} \\ \{x | h \geq 1.4\,\text{m or } h \leq 1\,\text{m}\}, & t \geq 1\,\text{s} \end{cases}$$

over the time range $[t_0, t_f] = [0, 5\,\text{s}]$, and a target set $T = \{x | h \in [0.98, 1.02]\}$ to be reached by time $t_f = 5\,\text{s}$. The set of initial states is the interval such that $p_n(0)$, $\dot{p}_n$ $p_e(0)$, $\dot{p}_e(0)$, $h(0)$, and $\dot{h}(0)$ all lie in $[-0.4, 0.4]$, and $\phi(0)$, $\dot{\phi}(0)$, $\theta(0)$, $\dot{\theta}(0)$, $\psi(0)$, and $\dot{\psi}(0)$ are all zero. To verify the safety and reachability specifications, we create a reachable tube approximation by computing reachable set over-approximations at 500 evenly spaced points in the range $[0, 5\,\text{s}]$, one of which is the point $t = 5\,\text{s}$ to verify that the target set is reached.

Figure 8.3 shows the reachable tubes computed for the quadrotor system benchmark, using the continuous-time mixed monotonicity from Sect. 4.1, sampled-data mixed monotonicity from Sect. 5.1, growth bound from Sects. 6.1 and 6.3, and probabilistic Monte Carlo methods from Sect. 7.2. For the sampled-data mixed-

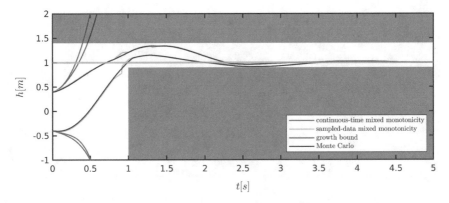

Fig. 8.3 Reachable tube approximations of the state h in the quadrotor model (8.1). To satisfy the design specifications, the trajectory of h must avoid the unsafe set (red) and reach the target height (green strip) by $t_f = 5\,\text{s}$

monotonicity method, the sensitivity matrix bounds were computed with the sampling technique of Sect. 5.3 using 500 samples. The Monte Carlo method used probabilistic parameters $\epsilon = 0.05$, $\delta = 10^{-9}$. On a computer with two physical cores at 2.6 GHz running MATLAB, computation took 32 s for continuous-time mixed monotonicity, 494 s for sampled-data mixed monotonicity, 19 s for growth bound, and 26 s for Monte Carlo. The sampled-data mixed-monotonicity and Monte Carlo methods both verify that the proposed control policy meets the specifications. However, the mixed-monotonicity and growth bound methods fail to verify the policy due to the conservativeness of their reachable set over-approximations. The conservativeness of these two methods quickly accumulates over time, unlike the approximations of the sampling-based methods. However, the sampling-based methods are not guaranteed to produce sound over-approximations (though the Monte Carlo method is guaranteed to be accurate in a probabilistic sense) and tend to be more computationally expensive.

References

Frehse G, Krogh BH, Rutenbar RA (2006) Verifying analog oscillator circuits using forward/backward abstraction refinement. In: Proceedings of the design automation & test in Europe conference, vol 1. IEEE

Immler F, Althoff M, Chen X, Fan C, Frehse G, Kochdumper N, Li Y, Mitra S, Tomar MS, Zamani M (2018) ARCH-COMP18 category report: continuous and hybrid systems with nonlinear dynamics. In: Proceedings of the 5th international workshop on applied verification for continuous and hybrid systems

Chapter 9
Measure of Robustness Against Parameter Uncertainty

This chapter applies interval reachability analysis to the problem of evaluating the robustness of a system against the effects of uncertain parameters. Specifically, we consider the volume of an interval reachable set as a continuous "performance measure" to gauge the robustness of a system, which can be used to guide the design of a robust controller. The volume of the interval over-approximation measures the effect of the parameter uncertainty on system trajectories starting in a given initial set: a smaller volume implies a smaller effect, and therefore, greater robustness. By treating the uncertain parameters as constant-valued inputs restricted to an interval range, computing this robustness measure reduces to solving Problems 1.1 and 1.2.

We illustrate this approach to robustness analysis on a medical exoskeleton model which has 6 states and 12 uncertain parameters. The dynamics of this system are too complicated to derive detailed system information such as bounds on the Jacobian matrix or growth bound functions. This precludes the use of the more efficient methods introduced in this book, so we must instead rely on the computationally expensive but widely applicable sampling-based methods from Sects. 5.3 and 7.2.

9.1 Medical Exoskeleton Model

The model used in this chapter was presented in Narvaez-Aroche et al. (2018). The exoskeleton and its user are modeled as a three-link planar robot with joints at the ankles, knees, and hips, as shown on the left side of Fig. 9.1. θ_1 is the angular position of link 1 (shanks) measured from the horizontal, θ_2 is the angular position of link 2 (thighs) relative to link 1, and θ_3 is the angular position of link 3 (torso) relative to link 2. The model has 12 parameters, which are the masses of the links m_1, m_2, and m_3; the moments of inertia about their respective centers of mass (CoMs) I_1, I_2, and I_3; their lengths l_1, l_2, and l_3; and the distances of their CoMs from the joints l_{c_1}, l_{c_2}, and l_{c_3}. The model has 4 inputs, which are a torque τ_h about the hips actuated

© The Author(s), under exclusive license to Springer Nature Switzerland AG 2021
P.-J. Meyer et al., *Interval Reachability Analysis*,
SpringerBriefs in Control, Automation and Robotics,
https://doi.org/10.1007/978-3-030-65110-7_9

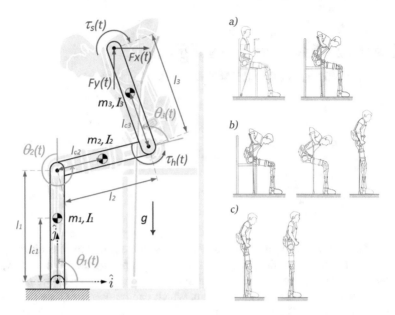

Fig. 9.1 *Left*: the medical exoskeleton and its user are modeled as a three-link planar robot. *Right*: phases of a dynamic sit-to-stand movement for a medical exoskeleton. **a** Preparation. **b** Ascension. **c** Stabilization. ©2020 IEEE. Reprinted, with permission, from Narvaez-Aroche et al. (2020)

by the exoskeleton, and a torque τ_s, and horizontal and vertical forces F_x and F_y, which capture the inertial and gravitational forces of the arms and loads applied on the shoulders of the user by their interaction with the ground through crutches.

In terms of the joint angles $\theta = [\theta_1; \quad \theta_2; \quad \theta_3]$, input $u = \begin{bmatrix} \tau_h; & \tau_s; & F_x; & F_y \end{bmatrix}$, and parameter vector

$$p = \begin{bmatrix} m_1; & m_2; & m_3; & I_1; & I_2; & I_3; & l_1; & l_2; & l_3; & l_{c_1}; & l_{c_2}; & l_{c_3} \end{bmatrix},$$

the Euler–Lagrange equations of the three-link planar robot can be written as

$$M\left(\theta\left(t\right),p\right)\ddot{\theta}\left(t\right) + F\left(\theta\left(t\right),\dot{\theta}\left(t\right),p\right) = A_\tau\left(\theta\left(t\right),p\right)u\left(t\right), \qquad (9.1)$$

where $M\left(\theta,p\right) \in \mathbb{R}^{3\times3}$ is the mass matrix of the system, $F\left(\theta,\dot{\theta},p\right) \in \mathbb{R}^3$ is the vector of energy contributions due to the acceleration of gravity and Coriolis forces, and $A_\tau\left(\theta,p\right) \in \mathbb{R}^{3\times4}$ is the generalized force matrix.[1] The states of the system are collected in the vector $x = \begin{bmatrix} \theta_1; & \theta_2; & \theta_3; & \dot{\theta}_1; & \dot{\theta}_2; & \dot{\theta}_3 \end{bmatrix}$. We consider a single initial state, $x_0 = [90°; \quad -90°; \quad 90°; \quad 0; \quad 0; \quad 0]$, which corresponds to a sitting position at rest.

---

[1]Describing the entries of $M\left(\theta,p\right)$, $F\left(\theta,\dot{\theta},p\right)$ and $A_\tau\left(\theta,p\right)$ is beyond the scope of this chapter. A full treatment of these matrices and their derivation is available in Narvaez-Aroche et al. (2017).

Table 9.1 Bounds $\left[\underline{p}, \overline{p}\right]$ for the parameter uncertainty of the system

Link	m_i (kg)	I_i (kg $\cdot$ m^2)	l_i (m)	l_{ci} (m)
1	[9.2, 10.2]	[1.10, 1.21]	[0.52, 0.54]	[0.23, 0.30]
2	[11.2, 13.2]	[0.49, 0.54]	[0.39, 0.42]	[0.17, 0.23]
3	[42.3, 46.8]	[2.40, 2.65]	[0.51, 0.53]	[0.24, 0.28]

The objective of the controller is to guide the state vector along a reference trajectory $\hat{x}(t)$ that brings the center of mass of the exoskeleton and its user from a sitting position to a standing position in a rest-to-rest maneuver, as shown on the right side of Fig. 9.1, between times $t_0 = 0$ and $t_f = 3.5$ s. The control policy is fixed as $u(t) = \hat{u}(t) + K_{LQR}(t) * (x(t) - \hat{x}(t))$, where $\hat{u}$ is a pre-computed reference input signal and K_{LQR} is an LQR gain matrix.

To account for the parameter uncertainty, the parameter vector p is treated as a constant input signal with values in the interval $\left[\underline{p}, \overline{p}\right]$, whose limits are shown in Table 9.1. This leads to a nonlinear system of the form (1.2) with $n_x = 6$ and $n_p = 12$.

9.2 Robustness Analysis with Sampling-Based Methods

The objective is to measure the performance and robustness to parameter uncertainty of the LQR controller designed above for the exoskeleton model. These measures are obtained by analyzing the volume of the reachable set approximations for the closed loop system. Due to the complicated nature of the terms in (9.1), the system details used in many of the reachability methods discussed in this book—Jacobian matrices, growth bounds, and Lipschitz constants—are not tractable to compute, and cannot be used in our analysis. We are, therefore, limited to methods that do not require these information, namely the sampling-based methods of Sects. 5.3 and 7.2.

To evaluate performance over the whole time range of the sit-to-stand motion, we approximate the reachable tube $\cup_{t\in[t_0,t_f]} R\left(t; t_0, x_0, \left[\underline{p}, \overline{p}\right]\right)$ by its sampled-time version at a 100 Hz sampling frequency. We thus compute a collection of 350 reachable sets: one every 10 ms over the time range [0, 3.5].

Analysis with Sampled-Data Mixed Monotonicity. The first considered reachability method corresponds to the sampled-data mixed monotonicity from Chap. 5, where bounds on the sensitivity matrix are evaluated through the sampling-based approach of Sect. 5.3. We first randomly draw a set of 500 parameter values within the interval $\left[\underline{p}, \overline{p}\right]$ using a Latin hypercube to ensure that these samples are well spread in the interval. For each sampled parameter, we compute a numerical evaluation of the trajectory of the sensitivity matrix over the whole time range [0, 3.5]. We then take the interval hull of all the sensitivity values evaluated at each of the 350 sampled times, and finally use these 350 sensitivity bounds to compute the corresponding reachable set approximations as in Proposition 5.1. We do not rely on the optional *falsification* step from Sect. 5.3 due to its too high computational cost on this system with 6 state variables and 12 parameters.

To compute this approximation of the reachable tube, the reachable set at each time $t \in [0, 3.5]$ is computed based on the evaluation of the sensitivity bounds $S_x \left(t; 0, x_0, \left[\underline{p}, \overline{p} \right] \right)$ at time t from an initial time $t_0 = 0$. Evaluating all sensitivity values with respect to the initial time $t_0 = 0$ has two main advantages compared to its alternative of evaluating the one-time-step evolution of the sensitivity based on the sensitivity values computed at the previous time step. The first advantage is that, it is sufficient to compute a single sensitivity trajectory over the whole time range $[0, 3.5]$ for each parameter value, instead of 350 sensitivity trajectories, each over a time range of 10 ms. The second advantage is that we avoid the accumulation of the approximation error since we do not reuse the possibly conservative results of the previous time step.

On a computer with four cores at 2.7 GHz running MATLAB Parallel Toolbox, this approach takes 63 min to obtain the numerical evaluation of the 500 trajectories of the sensitivity matrix and 115 min to compute the resulting sensitivity bounds and reachable set approximations for each of the 350 sampled times.

Analysis with Monte Carlo. The second reachability method is the Monte Carlo sampling-based approach of Sect. 7.2. This method consists of computing a set of sample trajectories from a set of parameters chosen uniformly at random over the interval $\left[\underline{p}, \overline{p} \right]$, and computing a reachable tube approximation by computing the componentwise minimum and maximum of the sample trajectories every 10 ms in the range $[0, 3.5]$ starting at 10 ms.

To determine the number of sample trajectories that need to be simulated, we choose an accuracy parameter ϵ and confidence parameter δ and select a large enough sample size to satisfy Proposition 7.2. Specifically, we choose $\epsilon = 0.05$ and $\delta = 10^{-9}$, which ensures that the reachable set approximation is at least 95% accurate (in the probabilistic sense discussed in Sect. 7.2), with at most a "one in a billion" chance that the approximation will fail to achieve this level of accuracy. According to the sample bound (7.1), this choice for ϵ and δ (and the state dimension $n_x = 6$) requires a sample size of $N = 1036$ samples in order for Proposition 7.2 to hold.

On a computer with two physical cores at 2.6 GHz running the MATLAB parallel computing toolbox, this method takes 134 min to simulate the system trajectories from the 1036 sampled parameters and use them to compute the Monte Carlo approximation for each of the 350 sampled times.

Discussion. For visualization, we focus our discussion of the results on a single state variable, namely θ_1. The θ_1-components of the reachable tube approximations computed using both sampling-based methods are shown on the left side of Fig. 9.2, along with an ensemble of trajectories with varying values for the parameter p.

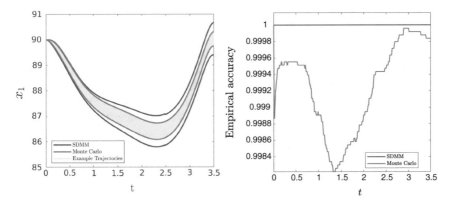

Fig. 9.2 *Left:* interval reachable tubes for state θ_1 computed by the sampled-data mixed-monotonicity (red) and Monte Carlo (blue) methods. System trajectories for varying values of p are also superimposed (yellow). *Right:* empirical estimates of the accuracy of the reachable set approximations, computed for each time t as the fraction of 24,700 sample trajectories that stay within the reachable set for time t

To experimentally verify that the assertion of Proposition 7.2 holds for the Monte Carlo approximations, and to validate the accuracy of the sampled-data mixed-monotonicity approximations, we compute a sample-based *empirical* estimate of the accuracy of the reachable set approximations computed by each method. Using a set of sample trajectories with parameters selected uniformly at random from $\left[\underline{p}, \overline{p}\right]$, we can compute the empirical accuracy of a reachable set approximation as

$$\text{empirical accuracy} = 1 - \frac{\text{number of trajectories that violate the approximation}}{\text{total number of trajectories}}.$$

For our estimate, we use 24,700 sample points. This sample size ensures that a *Chernoff bound*[2] holds, which guarantees that empirical accuracy is within 1% of the true accuracy (in the probabilistic sense of Sect. 7.2) with 99% confidence.

The empirical accuracy estimates for the θ_1-component of the reachable tubes are shown on the right side of Fig. 9.2. The empirical accuracy of the Monte Carlo approximation is 0.998 at its lowest. By applying the Chernoff bound, we conclude that the true probabilistic accuracy is no less than 0.988 (with 99% confidence), validating the claim of Proposition 7.2. On the other hand, the empirical accuracy of the sampled-data mixed-monotonicity approximation is exactly 1 for all reachable set approximations: the Chernoff bound ensures that the true accuracy is no less than 0.99, again with 99% confidence. This analysis shows that, while Monte Carlo sampling produces a tighter approximation, the sampled-data mixed-monotonicity method produces an estimate of greater probabilistic accuracy.

---

[2] A Chernoff bound is a probability inequality which provides a guarantee similar to Proposition 7.2 for a sum of random variables (the empirical accuracy in this case), though with a much higher sample complexity.

Further Reading

The medical exoskeleton model considered in this chapter was introduced in Narvaez-Aroche et al. (2017), and further developed and studied in Narvaez-Aroche et al. (2018, 2020); Devonport and Arcak (2020). The model of Narvaez-Aroche et al. (2018) is the one used in the numerical simulations shown above. The initial model employed feedback linearization-based controllers before adopting the LQR-based approach.

The use of interval volume as a performance metric to evaluate the robustness of the exoskeleton model was introduced in Narvaez-Aroche et al. (2020). Recall that the volume of a multidimensional interval $[a, b] \subset \mathbb{R}^{n_x}$ is

$$\text{vol}([a, b]) = \prod_{i=1}^{n_x} b_i - a_i.$$

Given an approximate reachable tube for the system, such as those computed in Sect. 9.2, the performance metric defined in Narvaez-Aroche et al. (2020) is

$$J = \sum_{t \in T_p} \text{vol}(r(t))$$

where $T_p \subset [t_0, t_f]$ is the finite set of time instants where the reachable set over-approximations are computed, and $r(t)$ is the interval reachable set over-approximation at time t. This performance metric can be used to compare the relative improvement in robustness between control policies. A lower value of J implies that a control policy is better able to reduce the impact of the parameter uncertainty on the system trajectory. In Narvaez-Aroche et al. (2020), this robustness performance metric is used to select a maximally robust LQR controller for the exoskeleton model by finding an LQR gain matrix which minimizes J.

References

Devonport A, Arcak M (2020) Estimating reachable sets with scenario optimization. In: Learning for dynamics and control

Narvaez-Aroche O, Packard A, Arcak M (2017) Motion planning of the sit to stand movement for powered lower limb orthoses. In: ASME 2017 dynamic systems and control conference. American Society of Mechanical Engineers Digital Collection

Narvaez-Aroche O, Meyer PJ, Arcak M, Packard A (2018) Reachability analysis for robustness evaluation of the sit-to-stand movement for powered lower limb orthoses. In: Dynamic systems and control conference, vol V001T07A006. American Society of Mechanical Engineers

Narvaez-Aroche O, Meyer PJ, Tu S, Packard A, Arcak M (2020) Robust control of the sit-to-stand movement for a powered lower limb orthosis. IEEE Trans Control Syst Technol 28(6):2390–2403

Chapter 10
Abstraction-Based Control Synthesis

Abstraction-based control synthesis is a three-step process to solve a control problem for continuous-state system (difference equation (1.1) or differential equation (1.2)), by first abstracting this system into a finite transition system, solving the control problem on the abstraction, and finally refining the obtained controller to apply it to the continuous system. Reachability analysis plays a key role in the first step to create the finite abstraction, as described in Sect. 10.1. This approach is then illustrated in Sect. 10.2 on a docking problem for a marine vessel. This example highlights how the *forward* reachability analysis methods presented in this book can be used within such abstraction-based approaches to solve *backward* reachability problems.

10.1 Abstraction-Based Control Using Interval Reachability Analysis

Assume that we are given a time-invariant system

$$\dot{x} = f(x, u, p), \tag{10.1}$$

with state $x \in X \subseteq \mathbb{R}^{n_x}$ evolving in the bounded interval $X \in \mathbb{I}^{n_x}$, bounded control input $u \in U \subseteq \mathbb{R}^{n_u}$, and bounded disturbance input $p \in P \subseteq \mathbb{R}^{n_p}$ in the disturbance interval $P \in \mathbb{I}^{n_p}$. An overview of the abstraction-based synthesis approach is sketched in Fig. 10.1. In Fig. 10.1a, we partition the state space into intervals and apply reachability analysis to the continuous system (from the gray partition cell, with control input u_1 in Fig. 10.1a) to obtain a finite transition system abstracting (10.1) in Fig. 10.1b. In Fig. 10.1c, we then synthesize a discrete controller on the abstraction to satisfy a given specification (avoiding the red state in Fig. 10.1c). Finally, in

© The Author(s), under exclusive license to Springer Nature Switzerland AG 2021   93
P.-J. Meyer et al., *Interval Reachability Analysis*,
SpringerBriefs in Control, Automation and Robotics,
https://doi.org/10.1007/978-3-030-65110-7_10

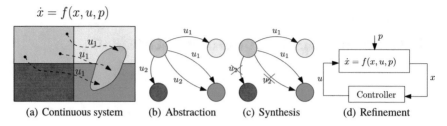

$$\dot{x} = f(x, u, p)$$

(a) Continuous system   (b) Abstraction   (c) Synthesis   (d) Refinement

Fig. 10.1 Abstraction-based synthesis for the control of a continuous system

Fig. 10.1d, we refine this discrete controller into a zero-order hold controller ensuring that (10.1) satisfies the same specification.

Abstraction. The first step is to abstract the continuous system (10.1) into a finite transition system defined as a triple $(\mathcal{X}, \mathcal{U}, \delta)$, where $\mathcal{X}$ is a finite set of states (also called *symbols*), $\mathcal{U}$ is a finite set of control inputs and $\delta : \mathcal{X} \times \mathcal{U} \to 2^{\mathcal{X}}$ is a nondeterministic transition relation describing the set of possible successors in $\mathcal{X}$ for each symbol-input pair.

To obtain this finite transition system $(\mathcal{X}, \mathcal{U}, \delta)$, we need to discretize all continuous sets and behaviors of (10.1). We first uniformly partition the interval state space $X \in \mathbb{I}^{n_x}$ into a finite set $\mathcal{X}$ of smaller identical intervals. The partition parameter is denoted as $\alpha_x \in \mathbb{N}$ and represents the number of elements per dimension, meaning that the whole partition $\mathcal{X}$ contains $\alpha_x^{n_x}$ intervals, each representing a discrete state of the abstraction. Next, we take a finite discretization $\mathcal{U}$ of the control set U. Increasing the number of elements in the discretization is beneficial for the control synthesis step, but increases the computational complexity of the abstraction step. The disturbance set P does not need to be discretized, since it will be fully handled within the reachability analysis, which is why the disturbance bounds P should be defined as an interval.

The last step of the abstraction is the discretization of the continuous behaviors of (10.1) to define the finite and nondeterministic transition relation $\delta : \mathcal{X} \times \mathcal{U} \to 2^{\mathcal{X}}$. Given a time sampling $\tau > 0$, we denote as $\hat{R}(\tau, \hat{x}, \hat{u}) \in \mathbb{I}^{n_x}$ the interval over-approximation of the reachable set $\{\Phi(\tau; x, \hat{u}, \mathbf{p}) \mid x \in \hat{x}, \mathbf{p} : [0, \tau] \to P\}$ of (10.1) starting from the partition element $\hat{x} \in \mathcal{X}$, with constant control input $\hat{u} \in \mathcal{U}$ and for any disturbance input function taking values in P. Then, the set of successors of the pair $(\hat{x}, \hat{u}) \in \mathcal{X} \times \mathcal{U}$ is defined as

$$\delta(\hat{x}, \hat{u}) = \left\{ \hat{x}' \in \mathcal{X} \mid \hat{x}' \cap \hat{R}(\tau, \hat{x}, \hat{u}) \neq \emptyset \right\}, \tag{10.2}$$

which corresponds to all partition elements in $\mathcal{X}$ with a non-empty intersection with the reachable set over-approximation. Using intervals in the partition and the over-approximations is thus beneficial not only for their low computational complexity, but also for their efficient implementation of the intersection tests in (10.2). Note that if X is not an invariant state space for (10.1), the set of states $\mathcal{X}$ of the finite abstraction needs to be expanded with one additional element Out representing the remaining state space $\mathbb{R}^{n_x} \setminus X$, from which no transition is created ($\delta(Out, \hat{u}) = \emptyset$), and for which all transitions ending in Out (whenever $\hat{R}(\tau, \hat{x}, \hat{u}) \not\subseteq X$) should be considered as unsafe in the control synthesis below.

Control Synthesis. Once the abstraction has been created, the second step is to synthesize a discrete controller on the abstraction to satisfy specifications. Working on a finite transition system allows us to consider a wide range of high-level specifications which can be defined as temporal logic formulas, and solved using existing graph search and model checking algorithms. In this chapter, we briefly discuss reachability specifications, which correspond to the control objective in the numerical example of Sect. 10.2. Details for other types of specifications are discussed in the *Further reading* section at the end of the chapter.

Given a target set $\mathscr{T} \subseteq \mathscr{X}$ contained in the finite set of abstract states, the reachability game with respect to $\mathscr{T}$ aims to find all abstract states that can be brought to $\mathscr{T}$ in a finite number of transitions of the abstraction. Solving this reachability game (where the controller plays against the nondeterminism of the abstraction) is achieved by first introducing the operator $G : 2^{\mathscr{X}} \rightarrow 2^{\mathscr{X}}$ defined for any $\mathscr{Z} \subseteq \mathscr{X}$ as

$$G(\mathscr{Z}) = \mathscr{T} \cup \left\{ \hat{x} \in \mathscr{X} \mid \exists \hat{u} \in \mathscr{U}, \, \delta(\hat{x}, \hat{u}) \subseteq \mathscr{Z} \right\}. \tag{10.3}$$

The fixed point of this operator initialized at the target set $\lim_{i \rightarrow \infty} G^i(\mathscr{T})$ is known to exist, to be reached in a finite number of iterations of G, and to be the largest subset of $\mathscr{X}$ solving this reachability game on the abstraction. The discrete controller associated with this winning set is synthesized at the same time by selecting the control inputs satisfying (10.3) at each new iteration of the operator G.

Controller Refinement. The third and last step of abstraction-based synthesis approaches is to refine the discrete controller into a controller for the continuous system (10.1). This is achieved by defining the zero-order hold version of the discrete controller with the chosen sampling period τ. This means that at each sampling instant, we measure the current continuous state, find the partition element (or abstract state) containing this continuous state, and apply the constant value of the discrete controller for this abstract state over the whole sampling period.

10.2   Reach–Avoid Problem for a Marine Vessel

In this section, we apply the abstraction-based approach from Sect. 10.1 to the kinematic model of a marine vessel:

$$\dot{x} = \begin{pmatrix} u_1 \cos(x_3) - u_2 \sin(x_3) + p_1 \\ u_1 \sin(x_3) + u_2 \cos(x_3) + p_2 \\ u_3 + p_3 \end{pmatrix}. \tag{10.4}$$

Fig. 10.2 States and control
inputs of a kinematic ship
model

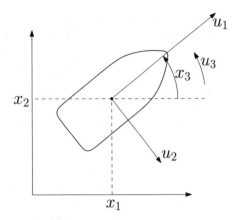

Fig. 10.2 States and control inputs of a kinematic ship model

As sketched in Fig. 10.2, the state $x \in \mathbb{R}^3$ describes the planar position (x_1, x_2) and the heading x_3 of the marine vessel, the control inputs $u \in \mathbb{R}^3$ are the surge velocity, sway velocity, and yaw rate, and the additive disturbance $p \in \mathbb{R}^3$ represents current velocities.

We consider a state space defined as the interval $X = [0, 10] \times [0, 6.5] \times [-\pi, \pi]$, with a target subset $X_r = [7, 10] \times [0, 6.5] \times [\pi/3, 2\pi/3]$ and two static obstacles $X_{a1} = [2, 2.5] \times [0, 3] \times [-\pi, \pi]$ and $X_{a2} = [5, 5.5] \times [3.5, 6.5] \times [-\pi, \pi]$. The control objective is to reach the target set X_r in finite time while staying within the safe space $X \backslash (X_{a1} \cup X_{a2})$. The input bounds are $U = [0, 0.18] \times [-0.05, 0.05] \times [-0.1, 0.1]$ for the control and $P = [-0.01, 0.01]^3$ for the disturbance.

Abstraction. To create an abstraction of (10.4), we consider a sampling time $\tau = 3$ s, a uniform partition of the state space X into 27000 smaller intervals ($\alpha_x = 30$ per dimension), and a uniform discretization of the input set U into 125 values ($\alpha_u = 5$ per dimension). The transition relation (10.2) is created using the continuous-time mixed monotonicity reachability method from Sect. 4.1. Due to the translational symmetry of (10.4) on its first two state variables, the whole transition relation δ can be created with only $\alpha_x \alpha_u^3 = 3750$ calls of the reachability analysis, instead of the $\alpha_x^3 \alpha_u^3 = 3.4$ millions calls that would have been required if the system had no such symmetry.

For each of these reachability analysis calls, we need to provide bounds on the non-constant terms of the Jacobian matrices

$$J_x(x, u) = \begin{pmatrix} 0 & 0 & -u_1 \sin(x_3) - u_2 \cos(x_3) \\ 0 & 0 & u_1 \cos(x_3) - u_2 \sin(x_3) \\ 0 & 0 & 0 \end{pmatrix}, \quad J_p = \begin{pmatrix} 1 & 0 & 0 \\ 0 & 1 & 0 \\ 0 & 0 & 1 \end{pmatrix}.$$

Although we could easily obtain Jacobian bounds satisfying Assumption 4.1 by globally bounding all sine and cosine functions by $[-1, 1]$, the dynamics of state x_3 in (10.4) actually allow us to compute local bounds for each call of the reachability analysis, and thus obtain less conservative over-approximations. Indeed, for each discrete transition, we have $\dot{x}_3 = u_3 + p_3$ with a constant control u_3 over the sampling period $[0, \tau]$, a bounded disturbance $p_3 \in \left[\underline{p_3}, \overline{p_3}\right] = [-0.01, 0.01]$, and an initial state $x_3(0) \in \left[\underline{x_{3,0}}, \overline{x_{3,0}}\right]$ starting in one of the partition elements. Then, possible values for x_3 over the sampling period $[0, \tau]$ are bounded as

$$x_3([0, \tau]) \in \left[\underline{x_{3,0}} + \min\left(0, \tau\left(u_3 + \underline{p_3}\right)\right), \overline{x_{3,0}} + \max\left(0, \tau\left(u_3 + \overline{p_3}\right)\right)\right].$$

Denoting this scalar interval as $\left[\underline{\chi_3}, \overline{\chi_3}\right]$ (which is different for each choice of $\left[\underline{x_{3,0}}, \overline{x_{3,0}}\right]$ and u_3), and since we know that $\left[\underline{\chi_3}, \overline{\chi_3}\right] \cap [-\pi, \pi] \neq \emptyset$ due to the initial states $\left[\underline{x_{3,0}}, \overline{x_{3,0}}\right] \subseteq [-\pi, \pi]$, then local extrema for the sine and cosine functions can be defined as follows:

$$\min_{x_3 \in \left[\underline{\chi_3}, \overline{\chi_3}\right]} \cos(x_3) = \begin{cases} -1 & \text{if } \{-\pi, \pi\} \cap \left[\underline{\chi_3}, \overline{\chi_3}\right] \neq \emptyset \\ \min\left(\cos\left(\underline{\chi_3}\right), \cos(\overline{\chi_3})\right) & \text{otherwise} \end{cases}$$

with similar equations replacing $\{-\pi, \pi\}$ by $\{-2\pi, 0, 2\pi\}$, $\left\{\frac{-5\pi}{2}, \frac{-\pi}{2}, \frac{3\pi}{2}\right\}$ and $\left\{\frac{-3\pi}{2}, \frac{\pi}{2}, \frac{5\pi}{2}\right\}$ for $\max(\cos(x_3)) = 1$, $\min(\sin(x_3)) = -1$ and $\max(\sin(x_3)) = 1$, respectively.

Control Synthesis. The control synthesis for the abstraction is done similarly to Sect. 10.1 with a fixed point on the operator G. Since we want to solve a reach–avoid specification and (10.3) only defines the solution to a backward reachability problem, we simply need to adapt the definition of the operator G in (10.3) by adding the condition $\hat{x} \cap (X_{a1} \cup X_{a2}) = \emptyset$. By taking several iterations of the call of G, this ensures that discrete states intersecting the obstacles are neither explored nor reached during the control synthesis.

Numerical Results. On a laptop with a 1.7 GHz CPU and 4 GB of RAM, the abstraction step took 45 s and the control synthesis 7.2 h. The large difference in computation times between these two steps is due to the fact that we can reduce the complexity of the abstraction creation by a factor of $\alpha_x^2 = 900$ by taking advantage of the translational symmetry of (10.4), but this cannot be exploited in the synthesis step. After the synthesis, the winning set for the reach–avoid specification covers 77% of the state space.

Figure 10.3 provides an example trajectory of the continuous system (10.4) controlled with a zero-order hold version of the discrete controller obtained on the abstraction. The system is initialized at $x_0 = [0; 0; 0]$ (thus starting in the bottom left corner and facing East), and the projection of the closed-loop trajectory in the x_1-x_2 plane is drawn in red, with the orientation x_3 represented by the blue arrows at each sampling time. We can then confirm that the controlled trajectory avoids both obstacles (in gray) and eventually reaches the target set corresponding to the 2D docking area in blue, with an orientation of the ship toward the North.

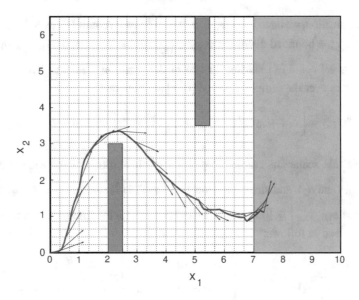

Fig. 10.3 Controlled trajectory (red) of the marine vessel (10.4) in the x_1-x_2 plane, with orientation x_3 (blue arrows), the target set (blue set) and the obstacles (gray sets)

To highlight the influence of the Jacobian bounds on these results and the interest of computing the local Jacobian bounds as above, we also applied this abstraction-based approach with all the same parameters as chosen above, except that the Jacobian bounds are computed based on the global bounds of the trigonometric functions: $\sin(x_3), \cos(x_3) \in [-1, 1]$. These larger bounds naturally lead to more conservative reachable set over-approximations, thus making the obtained abstraction less controllable. In particular, the added conservativeness implies that fewer new controllable states are found at each iteration of the control synthesis, which in turn, requires additional iterations to obtain the final controller. Although the winning set for the reach–avoid specification is only slightly smaller than previously (71% of the state space instead of 77%), the computation time is more than twice as much as before (15.1 h instead of 7.2) since a fixed point is only reached after 81 iterations of the synthesis (instead of 53 in the previous case).

Further Reading

This chapter illustrates how we can use the *forward* reachability analysis methods presented in this book to solve *backward* reachability problems, similar to the ones considered in the backward reachability tools relying on Hamilton–Jacobi–Isaacs equations (Mitchell et al. 2005).

Using reachable set over-approximations in the definition of the transition relation (10.2) guarantees a property called *alternating simulation relation* between the continuous model (10.1) and its finite abstraction ($\mathscr{X}$, $\mathscr{U}$, δ). This property ensures that any control strategy satisfying specifications on the abstraction can be refined into a control strategy for the continuous model (10.1) to satisfy the same specifications (Tabuada 2009). While the numerical example in Sect. 10.2 uses the reachability methods based on continuous-time mixed monotonicity from Sect. 4.1 to create an abstraction, there exist many other publications on abstraction-based synthesis relying on other interval reachability analysis methods to obtain the desired *alternating simulation relation*. These include monotonicity as in Chap. 3 (Moor and Raisch 2002), combining monotonicity with an internal feedback controller to reduce the conservativeness of the abstraction (Sinyakov and Girard 2020), a particular case of discrete-time mixed monotonicity with sign-stable Jacobian matrices (Coogan and Arcak 2015), and growth bounds (Girard et al. 2009; Reissig et al. 2016).

To fully define the transition relation (10.2), the number of required reachability analysis calls is exponential in the dimensions of the state space and control input space. Therefore, even for small systems, we can easily reach several millions of reachable sets needed for a single abstraction. This is the primary reason why only interval reachability analysis is used to create these abstractions, since other set representations would induce an even higher complexity. To further reduce the complexity of the abstraction creation for higher-dimensional systems, existing approaches include the use of a state space partition with varying granularity (Mouelhi et al. 2013; Hsu et al. 2018), or the decomposition of the continuous dynamics into subsystems (Reissig 2010; Boskos and Dimarogonas 2019; Kim et al. 2015; Dallal and Tabuada 2015; Pola et al. 2016; Meyer et al. 2017).

The nondeterministic abstractions created using reachability analysis as in this chapter are best suited for solving safety games (stay in a given set), reachability games (reach a target set in finite time), or combinations of them such as reach–avoid–stay specifications (reach a target while avoiding unsafe sets, then stay in the target). Although more complicated specifications (such as Linear Temporal Logic formulas (Baier and Katoen 2008)) can be handled by these abstractions (Belta et al. 2017; Coogan et al. 2017), the computational complexity can be daunting. One possible approach to address this problem is to decompose the high-level control problem into three layers of smaller problems easier to solve (model checking, trajectory search in the physical environment, and trajectory tracking on the dynamical system) (Meyer and Dimarogonas 2019). There also exist other abstraction-based approaches which are better suited to dealing with complex temporal logic specifications by creating a deterministic abstraction (Finucane et al. 2010; Tumova et al. 2010; Wongpiromsarn et al. 2011), at the cost of being limited to simpler dynamics (such as piecewise affine systems).

The numerical example in Sect. 10.2 focuses on the 3-dimensional kinematic model of a marine vessel. A more advanced approach is presented in Meyer et al. (2020), where the 6-dimensional dynamical model is initially reduced into its 3-dimensional kinematic model with guaranteed bounds on the error between the two models. The abstraction-based approach is then applied to the reduced model similarly to Sect. 10.2, but by considering a modified reach–avoid specification where the error bounds are used to shrink the target set and expand the obstacles, so that the satisfaction of this specification on the 3D model ensures the satisfaction of the main control objective on the 6D model.

References

Baier C, Katoen JP (2008) Principles of model checking. MIT press

Belta C, Yordanov B, Gol EA (2017) Formal methods for discrete-time dynamical systems, vol 89. Springer

Boskos D, Dimarogonas DV (2019) Decentralized abstractions for multi-agent systems under coupled constraints. Eur J Control 45:1–16

Coogan S, Arcak M (2015) Efficient finite abstraction of mixed monotone systems. In: 18th international conference on hybrid systems: computation and control, pp 58–67

Coogan S, Arcak M, Belta C (2017) Formal methods for control of traffic flow: automated control synthesis from finite state transition models. IEEE Control Syst Mag 37(2):109–128

Dallal E, Tabuada P (2015) On compositional symbolic controller synthesis inspired by small-gain theorems. In: IEEE conference on decision and control, pp 6133–6138

Finucane C, Jing G, Kress-Gazit H (2010) LTLMoP: experimenting with language, temporal logic and robot control. In: IEEE/RSJ international conference on intelligent robots and systems. IEEE, pp 1988–1993

Girard A, Pola G, Tabuada P (2009) Approximately bisimilar symbolic models for incrementally stable switched systems. IEEE Trans Autom Control 55(1):116–126

Hsu K, Majumdar R, Mallik K, Schmuck AK (2018) Multi-layered abstraction-based controller synthesis for continuous-time systems. In: Proceedings of the 21st international conference on hybrid systems: computation and control, pp 120–129

Kim ES, Arcak M, Seshia SA (2015) Compositional controller synthesis for vehicular traffic networks. In: IEEE conference on decision and control, pp 6165–6171

Meyer PJ, Dimarogonas DV (2019) Hierarchical decomposition of LTL synthesis problem for nonlinear control systems. IEEE transactions on automatic control 64(11):4676–4683

Meyer PJ, Girard A, Witrant E (2017) Compositional abstraction and safety synthesis using overlapping symbolic models. IEEE Trans Autom Control 63(6):1835–1841

Meyer PJ, Yin H, Brodtkorb AH, Arcak M, Sørensen A.J (2020) Continuous and discrete abstractions for planning, applied to ship docking. In: Proceedings of the 21st IFAC world congress (virtual), pp 1857–1862

Mitchell IM, Bayen AM, Tomlin CJ (2005) A time-dependent Hamilton-Jacobi formulation of reachable sets for continuous dynamic games. IEEE Trans Autom Control 50(7):947–957

Moor T, Raisch J (2002) Abstraction based supervisory controller synthesis for high order monotone continuous systems. In: Modelling, analysis, and design of hybrid systems. Springer, pp 247–265

Mouelhi S, Girard A, Gössler G (2013) CoSyMA: a tool for controller synthesis using multi-scale abstractions. In: Proceedings of the 16th international conference on hybrid systems: computation and control, pp 83–88

Pola G, Pepe P, Di Benedetto MD (2016) Symbolic models for networks of control systems. IEEE Trans Autom Control 61(11):3663–3668

Reissig G (2010) Abstraction based solution of complex attainability problems for decomposable continuous plants. In: IEEE conference on decision and control, pp 5911–5917

Reissig G, Weber A, Rungger M (2016) Feedback refinement relations for the synthesis of symbolic controllers. IEEE Trans Autom Control 62(4):1781–1796

Sinyakov V, Girard A (2020) Abstraction of monotone systems based on feedback controllers. In: 21st IFAC world congress

Tabuada P (2009) Verification and control of hybrid systems: a symbolic approach. Springer Science & Business Media

Tumova J, Yordanov B, Belta C, Černá I, Barnat J (2010) A symbolic approach to controlling piecewise affine systems. In: 49th IEEE conference on decision and control. IEEE, pp 4230–4235

Wongpiromsarn T, Topcu U, Ozay N, Xu H, Murray RM (2011) TuLiP: a software toolbox for receding horizon temporal logic planning. In: Proceedings of the 14th international conference on hybrid systems: computation and control. ACM, pp 313–314

Appendix A
Obtaining Sign-Stability by Shifting a Bounded Variable

This appendix shows how any bounded variable can be shifted to ensure that the range of values of the shifted variable is sign-stable, meaning that its values are either all nonnegative or all nonpositive. This method is particularly useful for the three reachability methods based on mixed monotonicity in Chaps. 4 and 5, and more precisely to compute the shifting matrices L_x and L_p in Assumptions 4.1, 4.2, and 5.1.

We consider a scalar variable $x \in \mathbb{R}$ and assume that its range of values is bounded on one side. This means that either there exists a lower bound $\underline{x} \in \mathbb{R}$ such that $x \in [\underline{x}, +\infty)$ or there exists an upper bound $\overline{x} \in \mathbb{R}$ such that $x \in (-\infty, \overline{x}]$. The particular case where the range of values of x is bounded on both sides corresponds to the overlap of both previous cases: there exist $\underline{x}, \overline{x} \in \mathbb{R}$ with $\underline{x} \leq \overline{x}$ such that $x \in [\underline{x}, \overline{x}]$.

We claim that this condition of one-side boundedness is equivalent to finding a constant scalar $y \in \mathbb{R}$ such that the shifted variable $x + y$ is sign-stable, which means that the range of values taken by $x + y$ is included either in the nonnegative half-plane $\mathbb{R}_+$ or in the nonpositive half-plane $\mathbb{R}_-$. Below, we thus show how the knowledge of the bounds on x can be used to constructively obtain the shifting constant y.

If x is Lower Bounded. Two separate cases are considered depending on the sign of the lower bound $\underline{x}$. For a positive lower bound as in Fig. A.1a, the range of values of x is already sign-stable since we have $[\underline{x}, +\infty) \subseteq \mathbb{R}_+$. Therefore, taking $y = 0$ is an acceptable choice since no shifting of this range is necessary. For a negative lower bound as in Fig. A.1b, taking $y = -\underline{x}$ implies that $\underline{x} + y = 0$ which ensures that the shifted variable only takes positive values: $x + y \in [0, +\infty)$. Therefore, the choice $y = -\underline{x}$ is the smallest possible shifting value guaranteeing the whole range of $x + y$ to become sign-stable. The two cases in Fig. A.1a, b using the lower bound can thus be summarized as

$$y = \max\left(0, -\underline{x}\right).$$

P.-J. Meyer et al., *Interval Reachability Analysis*,
SpringerBriefs in Control, Automation and Robotics,
https://doi.org/10.1007/978-3-030-65110-7

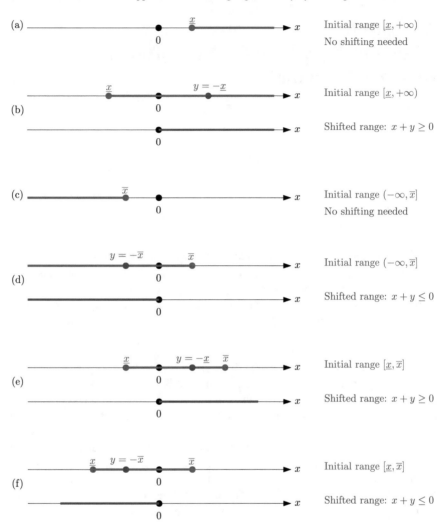

Fig. A.1 Illustrations on how a bounded interval $[\underline{x}, \overline{x}]$ (top of each subfigure) can be shifted to become sign-stable (bottom of each subfigure). Cases **a** and **b** for a lower-bounded interval, cases **c** and **d** for an upper-bounded interval, and cases **e** and **f** for an interval bounded on both sides

If x is Upper Bounded. We follow a similar reasoning as for the lower-bounded case. For a negative upper bound as in Fig. A.1c, the whole range $(-\infty, \overline{x}] \subseteq \mathbb{R}_-$ is already sign-stable in the negative half-plane, and it is thus sufficient to take $y = 0$. If the upper bound is positive as in Fig. A.1d, then $y = -\overline{x}$ is the smallest (in absolute value) value shifting the range to the negative half-plane. The two cases in Fig. A.1c, d using the upper bound can thus be summarized as

$$y = \min\left(0, -\overline{x}\right).$$

If x is Bounded on Both Sides. In this case, we can arbitrarily choose either of the two methods described in the previous paragraphs. It should be noted that when solving Problems 1.1 and 1.2 with reachability methods based on mixed monotonicity as in Chap. 4, the conservativeness of the obtained over-approximations is directly influenced by the absolute value of the shifting constant y. To minimize this conservativeness, it is thus advised to deal with the bound with lowest absolute value:

$$
y = \begin{cases} \max\left(0, -\underline{x}\right) & \text{if } \left|\underline{x}\right| \le \left|\overline{x}\right|, \\ \min\left(0, -\overline{x}\right) & \text{if } \left|\underline{x}\right| \ge \left|\overline{x}\right|. \end{cases}
$$

Indeed, this choice guarantees that the shifting distance $|y|$ is minimized in all cases. In particular, this ensures that $y = 0$ whenever the range $\left[\underline{x}, \overline{x}\right]$ is already sign-stable (similarly to Fig. A.1a, c).

For the non-scalar cases of Assumptions 4.1, 4.2 and 5.1 in Chaps. 4 and 5, our goal is to create the constant matrices $L_x \in \mathbb{R}^{n_x \times n_x}$ and $L_p \in \mathbb{R}^{n_x \times n_p}$ satisfying sign-stability conditions on the Jacobian matrices (Assumptions 4.1 and 4.2) or the sensitivity matrices (Assumption 5.1). Assuming that we are provided bounds for both Jacobian matrices (or the sensitivity matrix in the case of Assumption 5.1), then the shifting method described above in the scalar case can be applied independently for each element of these matrices. In the particular case of Assumption 4.1 in Sect. 4.1, there is no condition on the diagonal elements of the state Jacobian $J_x(t, x, p)$ and we can thus arbitrarily pick $L_{xii} = 0$ for all $i \in \{1, \ldots, n_x\}$.

The shifting matrices $L_x \in \mathbb{R}^{n_x \times n_x}$ and $L_p \in \mathbb{R}^{n_x \times n_p}$ obtained from this procedure are then guaranteed to satisfy Assumption 4.1, 4.2 or 5.1, depending whether we used bounds on the Jacobian matrices of the continuous-time system, bounds on the Jacobian matrices of the discrete-time system, or bounds on the sensitivity matrices of the continuous-time system, respectively.

Appendix B
TIRA Toolbox for Interval Reachability Analysis

This appendix gives an overview of *TIRA: Toolbox for Interval Reachability Analysis*, which is a MATLAB toolbox, initially published in [1], gathering several of the methods presented in this book, and is continually updated to include new ones. The toolbox and its documentation are publicly available at the following address:

https://gitlab.com/pj_meyer/TIRA

B.1 Toolbox Description

The development choices of TIRA were motivated by three main design goals: providing a library of several methods for interval reachability analysis; being easily extensible with new methods; and being usable by nonexpert users without requiring a detailed knowledge of the theory and implementation of these methods. To achieve these objectives, we designed the toolbox architecture so that each reachability method is implemented in its own function, and all methods can be accessed through a unique interface function. This interface function serves as a hub taking as inputs the definition of the reachability problem to be solved (time range and intervals of initial states and inputs), checking if the user-provided system satisfies the requirements of the requested reachability method, and finally calling this method and returning the resulting reachable set over-approximation.

This architecture greatly benefits the toolbox extensibility, since adding a new method only requires to implement this method in a separate function, then add a call for this function and the tests for its requirements in the main interface function. To allow its use by nonexpert users, it is also possible to call this interface function without specifying which reachability method to call. In such case, the interface function checks the requirements of all implemented methods and calls the most suitable one whose requirements are satisfied by the user-provided system. In addition,

© The Author(s), under exclusive license to Springer Nature Switzerland AG 2021
P.-J. Meyer et al., *Interval Reachability Analysis*,
SpringerBriefs in Control, Automation and Robotics,
https://doi.org/10.1007/978-3-030-65110-7

although we offer an easy access for the user to modify the internal parameters of
the reachability methods, if desired, all those parameters are predefined with default
values so that nonexpert users can still use the toolbox without the knowledge of
how each method and its internal parameters work.

The toolbox is implemented in MATLAB, and is thus platform independent and
does not require an installation. More details on requirements, setup, and use of the
toolbox are provided in the documentation available in the online repository at the
link in the introduction of this appendix.

B.2   Guided Use Through an Example

Several examples are already implemented in TIRA (folder Examples/) to test the
toolbox. In this section, we instead provide a didactic description of how to introduce
and solve a new reachability problem in TIRA. For this, we consider the tunnel diode
oscillator from Sect. 8.1.

The first step is to provide the system description by filling the skeleton function
System_description.m.

```
%% In file System_description.m

function dx = System_description(t,x,p)
% Constant parameters
C = 1e-12; L = 1e-6; R = 200; V_in = 0.3;
% Polynomial current-voltage relationship
i_D = @(v_D) 1e-3/.3*(803.712*v_D^5 - 1086.288*v_D^4 +
    551.088*v_D^3 - 124.548*v_D^2 + 10.656*v_D);
% Continuous-time dynamics
dx = [(-i_D(x(1))+x(2))/C; ...
      (-x(1)-R*x(2)+V_in)/L];
```

Depending on which reachability methods we want to apply to this system, we
might need to provide additional information on the system to satisfy the methods'
requirements. For this example, we first provide bounds on the Jacobian matrices in
file Input_files/UP_Jacobian_Bounds.m to satisfy Assumption 4.1 (with
the help of Appendix A) to use the method based on continuous-time mixed monotonicity in Sect. 4.1.

```
%% In file Input_files/UP_Jacobian_Bounds.m

function [J_x_low,J_x_up,J_p_low,J_p_up] =
    UP_Jacobian_Bounds(t_init,t_final,x_low,x_up,p_low,
    p_up)
% Constant parameters
C = 1e-12; L = 1e-6; R = 200;
% State Jacobian bounds
J_x_low = [-inf      1/C; ...
           -1/L      -R/L];
J_x_up = [.0074/C    1/C; ...
```

```
                 -1/L                 -R/L];
% Input Jacobian bounds
J_p_low = zeros(2); J_p_up = zeros(2);
```

We then provide in file `Input_files/UP_Contraction_Matrix.m` a contraction matrix satisfying Assumption 6.3 to use the growth bound method from Sect. 6.3.

```
%% In file Input_files/UP_Contraction_Matrix.m

function C = UP_Contraction_Matrix(t_init,t_final,
    x_low,x_up,p_low,p_up)
% Constant parameters
C = 1e-12; L = 1e-6; R = 200;
% Componentwise bounds on the state Jacobian matrix
C = [.0074/C      1/C; ...
     1/L          -R/L];
```

Another important file is `Solver_parameters.m`, in which all internal parameters of the toolbox are predefined and can be modified by the user. In particular, it can be useful to select a specific sub-method for one reachability analysis (e.g., how bounds on the sensitivity matrices are computed from Sects. 5.2–5.4, or how the growth bound function is created from Sects. 6.2–6.4). In this example, we modify the default parameters to ensure that the sensitivity bounds for the sampled-data mixed-monotonicity approach (Assumption 5.1 in Sect. 5.1) are computed based on the sampling method in Sect. 5.3 with 500 sample points and no falsification iteration. Note that the code below only includes the parameters in `Solver_parameters.m` whose default value was modified.

```
%% In file Solver_parameters.m

%% Sampled-data mixed monotonicity: method and
   parameter choice

% How are bounds on the sensitivity matrices obtained?
  % 1: Read sensitivity bounds from user-provided
    files
  % 2: Interval Arithmetics
  % 0: Sampling and falsification
parameters.sensitivity_bounds_method = 0;

%% Sampling and falsification: internal parameters

% Number of sample pairs in [x_low,x_up]*[p_low,p_up]
parameters.sampling_pairs = 500;
% Max number of falsification iterations
parameters.falsification_max_iter = 0;
```

The final steps are in the file `MAIN_CALL.m`, where we first define the reachability problem to be solved (intervals of initial states and inputs, and time range) and then call the toolbox for each desired reachability method. In this example, we call the three methods mentioned above: the continuous-time mixed monotonicity from

Sect. 4.1; the growth bound method from Sect. 6.1; and the sampled-data mixed-monotonicity method from Sect. 5.1 with the sampling-based sub-method from Sect. 5.3 as specified above in `Solver_parameters.m`. Although the numerical example in Sect. 8.1 approximates the reachable tube of this system by computing 500 reachable set over-approximations, the example code below only computes the reachable set over-approximations at the final time 15 ns.

```
%% In file MAIN_CALL.m

% Interval of initial states
x_low = [0.45; 0.6e-3];
x_up = [0.50; 0.6e-3];

% Interval of allowed input values
p_low = [0; 0];
p_up = [0; 0];

% Time interval
t_init = 0;
t_final = 15e-9;

% Toolbox call for continuous-time mixed-monotonicity
   method
[MM_low,MM_up] = TIRA([t_init,t_final],x_low,x_up,
   p_low,p_up,3);
% Toolbox call for growth bound method
[GB_low,GB_up] = TIRA([t_init,t_final],x_low,x_up,
   p_low,p_up,2);
% Toolbox call for sampled-data mixed-monotonicity
   method
[SD_low,SD_up] = TIRA([t_init,t_final],x_low,x_up,
   p_low,p_up,4);
```

Reference

Meyer PJ, Devonport A, Arcak M (2019) TIRA: toolbox for interval reachability analysis. In: Proceedings of the 22nd ACM international conference on hybrid systems: computation and control. ACM, pp 224–229

Series Editors' Biographies

Tamer Başar is with the University of Illinois at Urbana-Champaign, where he holds the academic positions of Swanlund Endowed Chair, Center for Advanced Study (CAS) Professor of Electrical and Computer Engineering, Professor at the Coordinated Science Laboratory, Professor at the Information Trust Institute, and Affiliate Professor of Mechanical Science and Engineering. He is also the Director of the Center for Advanced Study—a position he has been holding since 2014. At Illinois, he has also served as Interim Dean of Engineering (2018) and Interim Director of the Beckman Institute for Advanced Science and Technology (2008–2010). He received the B.S.E.E. degree from Robert College, Istanbul, and the M.S., M.Phil., and Ph.D. degrees from Yale University. He has published extensively in systems, control, communications, networks, optimization, learning, and dynamic games, including books on non-cooperative dynamic game theory, robust control, network security, wireless and communication networks, and stochastic networks, and has current research interests that address fundamental issues in these areas along with applications in multi-agent systems, energy systems, social networks, cyber-physical systems, and pricing in networks.

In addition to his editorial involvement with these Briefs, Başar is also the Editor of two Birkhäuser series on *Systems & Control: Foundations & Applications* and *Static & Dynamic Game Theory: Foundations & Applications*, the Managing Editor of the *Annals of the International Society of Dynamic Games* (ISDG), and member of editorial and advisory boards of several international journals in control, wireless networks, and applied mathematics. Notably, he was also the Editor-in-Chief of *Automatica* between 2004 and 2014. He has received several awards and recognitions over the years, among which are the Medal of Science of Turkey (1993); Bode Lecture Prize (2004) of IEEE CSS; Quazza Medal (2005) of IFAC; Bellman Control Heritage Award (2006) of AACC; Isaacs Award (2010) of ISDG; Control Systems Technical Field Award of IEEE (2014); and a number of international honorary doctorates and professorships. He is a member of the US National Academy of Engineering, a Life Fellow of IEEE, Fellow of IFAC, and Fellow of SIAM. He has served as an IFAC Advisor (2017–), a Council Member of IFAC (2011–2014),

© The Author(s), under exclusive license to Springer Nature Switzerland AG 2021
P.-J. Meyer et al., *Interval Reachability Analysis*,
SpringerBriefs in Control, Automation and Robotics,
https://doi.org/10.1007/978-3-030-65110-7

president of AACC (2010–2011), president of CSS (2000), and founding president of ISDG (1990–1994).

Miroslav Krstic is Distinguished Professor of Mechanical and Aerospace Engineering, holds the Alspach endowed chair, and is the founding director of the Cymer Center for Control Systems and Dynamics at UC San Diego. He also serves as Senior Associate Vice Chancellor for Research at UCSD. As a graduate student, Krstic won the UC Santa Barbara best dissertation award and student best paper awards at CDC and ACC. Krstic has been elected Fellow of IEEE, IFAC, ASME, SIAM, AAAS, IET (UK), AIAA (Assoc. Fellow), and as a foreign member of the Serbian Academy of Sciences and Arts and of the Academy of Engineering of Serbia. He has received the SIAM Reid Prize, ASME Oldenburger Medal, Nyquist Lecture Prize, Paynter Outstanding Investigator Award, Ragazzini Education Award, IFAC Nonlinear Control Systems Award, Chestnut textbook prize, Control Systems Society Distinguished Member Award, the PECASE, NSF Career, and ONR Young Investigator awards, the Schuck ('96 and '19) and Axelby paper prizes, and the first UCSD Research Award given to an engineer. Krstic has also been awarded the Springer Visiting Professorship at UC Berkeley, the Distinguished Visiting Fellowship of the Royal Academy of Engineering, and the Invitation Fellowship of the Japan Society for the Promotion of Science. He serves as Editor-in-Chief of *Systems & Control Letters* and has been serving as Senior Editor for *Automatica* and *IEEE Transactions on Automatic Control*, as editor of two Springer book series—Communications and Control Engineering and *SpringerBriefs in Control, Automation and Robotics*—and has served as Vice President for Technical Activities of the IEEE Control Systems Society and as chair of the IEEE CSS Fellow Committee. Krstic has coauthored thirteen books on adaptive, nonlinear, and stochastic control, extremum seeking, control of PDE systems including turbulent flows, and control of delay systems.

Printed in the United States
By Bookmasters